TRADE PESSIMISM AND REGIONALISM IN AFRICAN COUNTRIES: THE CASE OF GROUNDNUT EXPORTERS

Ousmane Badiane
Sambouh Kinteh

Research Report 97
International Food Policy Research Institute
Washington, D.C.

Library of Congress Cataloging-
in-Publication Data

Badiane, Ousmane.
    Trade pessimism and regionalism in African
countries : the case of groundnut exporters / by
Ousmane Badiane and Sambouh Kinteh.
        p.   cm. — (Research report ; 97)
    Includes bibliographical references and index.
    ISBN 0-89629-100-6
    1. Peanut industry—Africa. 2. Oil industries—
Africa. I. Kinteh, Sambouh. II. Title. III. Series:
Research report (International Food Policy Research
Institute) ; 97.

HD9235.P32A353   1994   94-7801
338.1′756596′096—dc20   CIP

CONTENTS

Foreword   vii

1. Summary   1

2. Introduction   3

3. Trends in World and African Groundnut Council Production
   of Groundnuts and Other Oilseeds   9

4. Trends in World and African Groundnut Council Oilseed
   Trade   14

5. External Demand Constraint, Domestic Policies, and Export
   Performance of African Groundnut Council Countries   21

6. The Groundnut Demand Outlook and the Potential Role of
   Regional Markets   51

7. Conclusions   69

Appendix 1: Findings and Recommendations of a Joint Study by
            the United Nations Economic Commission for Af-
            rica and the Food and Agriculture Organization of
            the United Nations   72

Appendix 2: Estimation of the Equilibrium Exchange Rate   74

Appendix 3: The Constant Market Share Model   76

References   78

TABLES

1. World and African production of 11 major oilseeds and oleaginous fruits, 1961-88   9

2. Production, area, and yield of groundnuts in shell, by selected regions and countries, 1961-88   11

3. Production of major oilseeds and oleaginous fruits other than groundnuts in AGC countries, 1961-88   12

4. Production, area harvested, and yield, in shell, of African groundnut production, selected countries, 1961-88   13

5. World exports of 10 major oilseeds and oleaginous fruit products, 1961-87   14

6. World exports of groundnut products by regions and selected countries, 1961-87   16

7. World imports of groundnut products by region and selected countries, 1961-87   17

8. African and AGC exports of groundnuts and oil palm products, 1961-87   18

9. Imports of major oilseeds and oleaginous fruit products by selected African countries   20

10. Groundnut oil trade and production performance, AGC countries compared with selected regions and the world, 1961-65 and 1986-88   21

11. Decomposition of annual changes in the real value of groundnut exports, 1961-87   26

12. Decomposition of the changes in real groundnut producer prices, 1968-88   31

13. Direct effects of sector policies on real country groundnut prices, The Gambia, Senegal, and Sudan, 1966-88   39

14. Ratios of producer prices and transfer costs to groundnut export prices, The Gambia, Senegal, and Sudan, 1966-88   40

15. Exchange rate disequilibrium in The Gambia, Senegal, and Sudan, 1966-88   42

16. Direct and aggregate effects of country policies on groundnut prices in The Gambia, Senegal, and Sudan, 1966-88   44

17. Estimated average annual change in groundnut output as a result of policies in AGC countries, 1967-88   48

18. Effects of domestic policies on groundnut exports, 1967-88   49

19. Estimated average annual change in aggregate groundnut exports as a result of AGC country policies, 1967-88   50

20. World imports of major oilseeds and oleaginous fruit products, by region, 1961-87   52

21. Share of vegetable oils and fats in daily calorie intake, selected countries, 1979-81   54

22. Trends in per capita consumption of major fats and oils, selected countries, 1960-87   55

23. Projected groundnut imports by economic regions   56

24. Market share results for oilseed and oilfruit products, AGC countries, 1962-87   58

25. Groundnut and palm oil imports and groundnut oil exports by West African and other African countries, 1963-87   59

26. Estimates of demand parameters for regional oilseed imports   66

27. Estimates of demand parameters for groundnut imports from AGC and non-AGC sources   66

ILLUSTRATIONS

1. Income terms of trade for groundnuts, The Gambia, 1961-87   4

2. Income terms of trade for groundnuts, Mali, 1961-87   4

3. Income terms of trade for groundnuts, Senegal, 1961-87   5

4. Income terms of trade for groundnuts, Sudan, 1961-87   5

5. World prices of groundnuts compared with soybeans and palm, 1962-87   24

6. World prices of groundnuts compared with sunflower seed and rapeseed, 1962-87   24

7. World exports of oilseeds, 1961-87   25

8. Non-AGC and AGC exports of groundnuts, 1961-87   25

9. Direct nominal protection of groundnuts in selected AGC countries, 1966-88   38

10. Degree of divergence from the equilibrium exchange rate   42

11. Total nominal protection of groundnuts in The Gambia, Senegal, and Sudan, 1966-88   43

12. Short-term divergence between actual and equilibrium levels of groundnut output, 1967-88   46

13. Long-term divergence between actual and equilibrium levels of groundnut output, 1967-88   47

14. Short-term divergence between actual and equilibrium levels of groundnut export revenue, 1967-88   47

15. Long-term divergence between actual and equilibrium levels of groundnut export revenues, 1967-88   48

16. Share of Africa in world imports of palm oil and groundnut oil, 1963-87   60

17. Export share of selected AGC countries in groundnut oil, 1961-87   60

18. Changes in groundnut oil imports between 1963 and 1987   61

19. Changes in palm oil imports between 1963 and 1987   62

FOREWORD

Because of the importance of agricultural trade for sustained growth in the agricultural sector and for poverty alleviation among African countries, research in this area has been a key component of IFPRI's research program in that continent. The present report by Ousmane Badiane of IFPRI and Sambouh Kinteh of the African Groundnut Council, based in Lagos, Nigeria, is the outgrowth of a research network on regional agricultural trade in West Africa initiated by IFPRI in 1989. It uses the example of the groundnut sector to examine the relationship between the domestic policy environment surrounding the agricultural sector and the performance of exports on traditional and regional export markets.

After long periods of strong performance in global markets, primary exports from many African countries have declined rapidly since the 1970s. African policymakers and analysts have reacted to this weakening performance with increasing pessimism regarding the long-term contribution of traditional export markets. They are stressing, instead, the importance of expanding trade in intra-African markets as an alternative to traditional export markets.

In examining the pitfalls of export pessimism and the limits of regionalism in trade by African countries, the report finds that African exporters have been more vulnerable to changes in international markets than competing exporters largely because domestic sectoral and macroeconomic policies have reduced their competitiveness. Regional markets could indeed play a significant role in African oilseed trade and market shares could be maintained or expanded, the report concludes, if exporters were able to cut unit costs in production, processing, and trading.

This places domestic factors that shape the conditions for production and export in African countries at the core of the problem. If African countries succeed in eliminating the internal obstacles and disincentives facing their exporters, then there will be less ground for pessimism. Alternatively, if African countries fail to adequately address the role of domestic factors in the performance of their trading sectors, they will be unlikely to take any real advantage of the proximity of regional markets. This is an important lesson likely to be relevant not only for groundnuts in West Africa but for efforts to expand regional agricultural trade more generally.

Per Pinstrup-Andersen
Director General

ACKNOWLEDGMENTS

The authors thank the Swiss Development Corporation for financial support. We are also grateful to Jayashree Sil for her excellent research assistance.

1

SUMMARY

The contribution of groundnut production, processing, and trade to the development of the African Groundnut Council (AGC) member economies during the first decades after their independence has been vital. Until the mid-1970s, the groundnut sector accounted for a large share of gross domestic product (GDP) and was the main source of export revenue and rural employment in AGC countries (The Gambia, Mali, Niger, Nigeria, Senegal, and Sudan). Between 1961 and 1965, 25 percent of the world's groundnut production took place in AGC countries. During the same period, AGC countries had a 62 percent share of world exports of groundnut oil. Although the countries are still major exporters, when taken collectively, and groundnuts are still a major sector in most of the member economies, their role in international markets, and that of groundnuts in their economies, has changed dramatically since the end of the 1970s. The changes in the groundnut sector are reflected in a rapid decline of AGC-wide production and exports.

Adverse developments on international markets, such as falling global demand and declining world market prices, are the most commonly cited reasons for the problems plaguing the groundnut sector in AGC countries. The growing pessimism about the long-term development of overseas export markets induced officials in member countries to adopt the Banjul Plan of Action for Groundnuts. The plan strongly emphasizes the promotion of intra-African trade and the recapturing of regional import markets as a component of the rehabilitation strategy for the groundnut sector.

This report investigates the concerns related to developments on international markets, on the one hand, and the potential of regional markets to contribute to the rehabilitation of the groundnut industry in AGC countries, on the other hand. The sources of the decline of groundnut trade in AGC countries are analyzed using data from The Gambia, Senegal, and Sudan—the member countries that have consistently exported throughout the study period. Contrary to the argument of an external demand constraint, the data reveal a much stronger contribution by domestic policies than by changes on global markets to the export performance of individual member countries.

Despite stable world demand for groundnut oil, AGC exports have fallen by more than two-thirds since the late 1960s, while other developing Asian and South American exporters have nearly quadrupled their market shares. Furthermore, the results of the decomposition of the changes in real export values indicate that changes on the supply side have played the predominant role in the decline of AGC groundnut trade.

Domestic sector and macroeconomic policies in AGC countries have played a particularly critical role in the decline of production and exports by AGC countries. Their combined effect on national groundnut sectors has resulted in average annual changes in AGC-wide groundnut production of -4 to -15 percent from the late 1960s to the end of the 1980s. The induced average annual decline in total AGC groundnut exports ranges from 17 to 40 percent, with the strongest decline during the 1970s.

1

The report also examines the potential of regional markets to contribute to AGC groundnut exports by investigating, first, the extent to which member countries have been able to take advantage of the geographic proximity of these markets and, second, the main determinants of import demand for various vegetable oils at the regional level. The results show that regional markets have hardly played a role in AGC exports. The region's share in the exports of the AGC's largest exporting country, Senegal, has been as low as 3 percent, although demand in these markets has expanded at a rate two-and-a-half times higher than demand in world markets. AGC exporters do not seem to have gained any real advantage from their proximity to regional markets.

However, the analysis of the determinants of regional oilseed import demand suggests that, even with stagnant demand, AGC exporters could increase both the quantity and value of their exports to regional markets and raise their market share by cutting the costs of production and distribution in order to contain competition from non-AGC exporters. The hope expressed at the Banjul meeting that intensifying regional trade in groundnut products would help solve the problem faced by AGC exporters cannot be realized unless appropriate changes in domestic policies that affect groundnut production and trade take place. AGC exports to the region and elsewhere have suffered much more from domestic policies than from external demand constraints.

2

INTRODUCTION

Background

From independence to the beginning of a major Sahelian drought in 1969, a central feature of the overall development experience in the African Groundnut Council (AGC) countries has been the overriding role played by the export-oriented groundnut sector in these economies. The countries that are members of the AGC are The Gambia, Mali, Niger, Nigeria, Senegal, and Sudan. Well before the countries became independent, groundnut production, marketing, and trade served as major sources of employment, income, and foreign exchange. The groundnut sector not only provided the basis for agro-industrial development, but it contributed significantly to the commercialization, monetization, and integration of the national rural sectors.

Although the contributions of groundnuts varied markedly among countries and within any one country over time, the types of growth linkages they generated between agriculture and nonagriculture and between the domestic economy and external markets were extensive. Hence, the groundnut industry was the dominant force affecting the pace, stability, and robustness of growth. The following examples illustrate the importance of the groundnut sector to the economies of AGC countries.

Until the mid-1970s, the groundnut sector contributed 30-40 percent of the gross domestic product (GDP) of The Gambia (UNECA/FAO 1985). Before the inception of the Sahelian drought in the late 1960s, about 12 percent of the GDP in Niger originated from the groundnut sector (Abdoulaye 1988). Similarly, 14 percent of GDP in Senegal was produced in that sector during the period between the droughts of 1970 and 1977 (UNECA/FAO 1985).

Before the mid-1970s, exports of groundnut products were a major source of foreign exchange for AGC countries, accounting for 40-90 percent of export revenues in all AGC member countries except Nigeria and Sudan, where the share of groundnuts in total export revenues in the late 1960s and early 1970s was close to 25 percent (Agbola and Opadokun 1984; Drave and Dembele 1984; El Bashir and Idris 1983; UNECA/FAO 1985; Issaka 1984).

The contribution to employment of the groundnut industry in these countries, with its ramifications for social and political stability, was even more substantial. Groundnut production is the main activity in The Gambia's agricultural sector. Until recently, it provided employment for as much as 80 percent of the Gambian population. In the late 1960s groundnut production employed about 30 percent of the rural population of Niger. Before the early 1980s, groundnut production, processing, and marketing provided employment for 70 percent of the active rural population in Senegal and occupied about 50 percent of the total cultivated land annually (UNECA/FAO 1985; Issaka 1984; Agbola and Opadokun 1984).

The contribution of the groundnut sector, particularly groundnut trade, to national economies changed dramatically after the mid-1970s. Figures 1-4 show the evolution

Figure 1—Income terms of trade for groundnuts, The Gambia, 1961-87

Index
(1961=100)

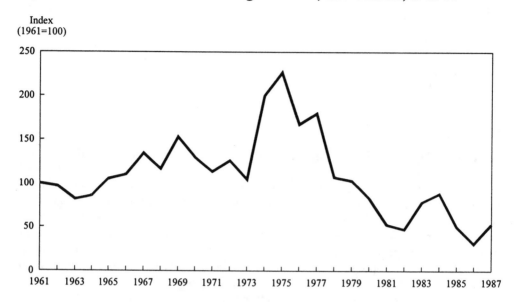

Source: Export revenue is from Food and Agriculture Organization of the United Nations, *FAO Trade Yearbook* (Rome: FAO, various years).
Note: Groundnut export revenue is deflated by the world manufacturing unit value.

Figure 2—Income terms of trade for groundnuts, Mali, 1961-87

Index
(1961=100)

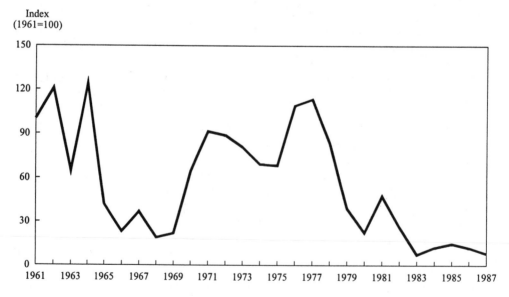

Source: Export revenue is from Food and Agriculture Organization of the United Nations, *FAO Trade Yearbook* (Rome: FAO, various years).
Note: Groundnut export revenue is deflated by the world manufacturing unit value.

Figure 3—Income terms of trade for groundnuts, Senegal, 1961-87

Index
(1961=100)

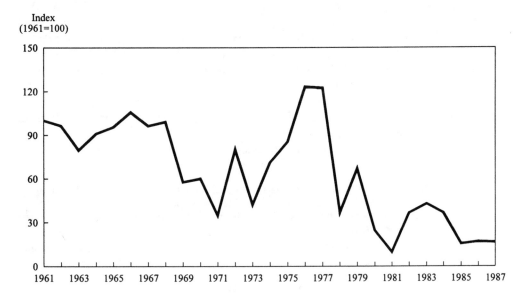

Source: Export revenue is from Food and Agriculture Organization of the United Nations, *FAO Trade Yearbook*
(Rome: FAO, various years).
Note: Groundnut export revenue is deflated by the world manufacturing unit value.

Figure 4—Income terms of trade for groundnuts, Sudan, 1961-87

Index
(1961=100)

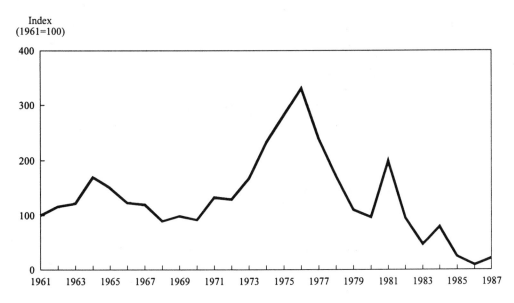

Source: Export revenue is from Food and Agriculture Organization of the United Nations, *FAO Trade Yearbook*
(Rome: FAO, various years).
Note: Groundnut export revenue is deflated by the world manufacturing unit value.

of groundnut export income terms of trade for several AGC countries, calculated as groundnut export revenues deflated by the world manufacturing unit value (MUV). After a period of relative stability during the 1960s, the income terms of trade fell continuously over the next one-and-a-half decades. Toward the end of the 1980s the real value of groundnut export revenues was about 50-80 percent lower than in the 1960s. The trends depicted in Figure 1 are similar across AGC countries; that is, pessimism is strong about the long-term contribution of the external sector to growth in the groundnut economy. The reasons are still the subject of a great deal of speculation, however, not withstanding the tendency for national policymakers to look to external factors for explanation.

The following excerpts from the introductory note to the Banjul meeting by the AGC's executive secretary illustrate the prevailing pessimism.

> The initiative of the Council of Ministers [to organize the Banjul meeting] was based on the conviction that our European market for groundnut and its products was steadily shrinking, as a result of the competition from other edible oils of various origins, including colza, soya, and sunflower (AGC 1984, 1).

> This symposium is being held at a critical period, when economic recession in major industrial countries is adversely affecting the primary commodity exports of nonpetroleum developing countries, just as it did in 1975. For this reason, some critical observers fear that the impact of two opposing economic forces might bring about a fresh crisis on the world market and consequently undermine at the international level the already fragile position of heavily indebted developing countries (AGC 1984, 2).

> Meanwhile, we should bear in mind the need to be self-reliant. Therefore, consideration should be given to ways and means of intensifying intra-African trade in groundnut and groundnut products, in spite of all obstacles that may stand in the way (AGC 1984, 3).

The recommendations arrived at in Banjul also reflect the concern about international markets and the role of regional markets:

> In addition, the possibility of expanding intra-African trade could be enhanced and, to this end, the following steps should be taken:

> i. The State Trading Organization of interested African countries should initiate regular contacts to examine the possibility of increasing trade among them.

> ii. The elimination and/or reduction of tariff and nontariff barriers in African countries should be sought, and the existing sub-regional groupings such as ECOWAS [Economic Community of West African States], CEAO [Communauté Economique de l'Afrique de l'Ouest], and UDEAC [Union Douanière et Economique de l'Afrique Centrale] are possible for initiating discussion on these aspects.

> iii. A comprehensive study should be undertaken to identify and survey potential import as well as supply markets in Africa. In this context, specific points that should be examined include methods of settlement of payment through bodies like the West African Clearing House, transportation links, the possibility of expanding these markets through promotion campaign.

> The implementation of the recommendations concerning groundnut production and marketing should be pursued vigorously, bearing in mind that the attainment of self-sufficiency in food production and the expansion of intra-African trade are among the main priority objectives from the economic development of Africa as embodied in the Lagos Plan of Action (AGC 1984, 442).[1]

---

[1] A less enthusiastic view of the potential role of intraregional trade can be found in UNECA/FAO 1985, 86.

Following the Banjul meeting, the Council requested that a first study be carried out "on groundnut production, marketing, processing and trade situation in AGC-member countries" by the United Nations Economic Commission for Africa (UNECA). (See Appendix 1 for the terms of reference and recommendations.) That study, completed in 1985, focused primarily on nonprice factors such as seed production and accessibility (subsidy), fertilizer delivery, and processing (UNECA/FAO 1985). The report blames the decline of the groundnut sector in individual countries on drought, disease, and extension policies.

Although the report offers a good overview of the problems, it does not carry the analysis of price factors far enough, nor does it analyze the demand for groundnuts and other oilseeds on regional markets. Moreover, low producer prices are seen as a problem primarily in Niger and Mali. On the marketing side, the analysis centers around the difficulties faced by state marketing boards in carrying out their procurement and stabilization activities.

Regarding the issue of raising AGC groundnut exports to regional markets, the report concludes with the following recommendation:

> Intra-African trade in groundnut products has not been significant due, partly, to the existence of tariff and nontariff barriers to groundnut trade in Africa. It is, therefore, recommended that these barriers should be removed forthwith in order to promote intra-African trade in groundnut products as well as in other oilseeds (UNECA/FAO 1985, 102).

In this report by the International Food Policy Research Institute, which was carried out in collaboration with AGC's Executive Secretariat, stronger emphasis is placed on the impact of price factors on the performance of member countries in the production of and trade in groundnut products. Particular attention is given to the role of groundnut pricing and marketing policies as well as to the overall economic policies of individual countries that discourage groundnut production and trade, including exports to regional markets. The determinants of regional demand for groundnut and other oilseed imports are analyzed to identify the factors that determine the ability of AGC countries to raise exports to these markets and profit from the expansion of demand in intra-African markets.

Objectives

Against this background, the study pursues two main objectives. First, to test the validity of the frequently made argument that external demand stagnation is responsible for the declining contribution of the groundnut sector to AGC member countries' economies. This is done by evaluating the role played by changes in foreign demand and by internal factors, particularly those in the domestic groundnut sector and macroeconomic policy environments.

Given the general emphasis of policy debates and of the Banjul conference on intra-African trade, the second major objective of the study is to look into the possibility of AGC countries taking advantage of import demand on regional markets in their strategies to revitalize their groundnut export sectors.

Chapters 3 and 4 analyze the trends in production and trade of groundnuts and other oilseeds in AGC countries and the world. The focus is on the changes in global flows of groundnut products relative to other oilseeds and on the relative performance of individual AGC exporters, compared with other groundnut exporting countries. The world oilseeds economy has gone through tremendous changes since the 1960s. One main feature of these changes is the surge of soybean and palm oil production in Asia and South America and of rapeseed and sunflower seed production in the European Community (EC). The other major change is the shift in the product forms in which the various oilseeds are traded. Related to these two types of changes is the emergence of new suppliers and the shift in the position of traditional exporters in international markets. Disentangling the various influences emanating from these changes is a necessary step toward understanding the development of the AGC countries' trade performance.

In Chapter 5 an attempt is made to isolate the contributions of domestic and external factors to the changes in each member country's export performance. The main concern in this part of the study is to shed some light on the controversy surrounding the external demand constraint argument. The critical importance of country sectoral and macroeconomic policies in coping with developments on international markets and exploiting the potential for regional markets is also shown, an element that does not always receive the attention it deserves. The emphasis on domestic policies is necessary to show their role in the decline of the groundnut sector and to tie the strategic debates on groundnuts back to the economic environment in member countries.

The analysis in Chapter 6 turns to the long-term prospects of groundnut demand on global and regional markets. First, the dynamics of the world demand for vegetable oils are debated in order to identify eventual future markets for AGC exporters. The past history of regional markets and their potential as future outlets are investigated, starting with a study of the extent to which AGC countries have taken advantage of the geographic proximity of regional markets. The main determinants driving vegetable oil import demand at the regional level are investigated. Finally, conclusions drawn from the report are reported in Chapter 7.

3

TRENDS IN WORLD AND AFRICAN GROUNDNUT COUNCIL PRODUCTION OF GROUNDNUTS AND OTHER OILSEEDS

During the three decades from the 1960s through the 1980s, world production of oilseed products was characterized by rapid technological change, which produced dramatic shifts in the location of production and in the relative importance of individual vegetable oils and fats (Table 1). During the period 1982-87 soybean production was dominant, with an annual average output of 93 million metric tons,[2] followed by coconut, cottonseed, groundnuts, sunflower seed, and palm oil, which had production levels ranging from 7 to 38 million tons. The remaining oilseeds are much less important, with production volumes of less than 5 million tons.

Looking at production growth rates, however, a different ranking emerges. Palm oil, with an annual production growth rate of 8 percent during 1961-88, leads the group, followed by soybean, sunflower seed, and palm kernel oil, with growth rates between 4.0 and 5.5 percent.[3] The most dramatic changes occurred in palm oil,

Table 1— World and African production of 11 major oilseeds and oleaginous fruits, 1961-88

Commodity	World		Africa	
	1982-87 Average	1961-88 Annual Growth Rate	1982-87 Average	1961-88 Annual Growth Rate
	(million metric tons)	(percent)	(1,000 metric tons)	(percent)
Groundnuts	20.08	1.31	4,151.71	−1.10
Soybeans	92.93	5.51	371.81	8.29
Coconuts	37.52	1.98	657.02	−0.89
Palm oil	7.35	7.82	208.59	1.79
Palm kernels	2.42	3.99	2,191.94	0.76
Castor beans	1.01	1.34	37.59	−2.57
Sunflower seed	18.17	4.04	452.84	7.61
Sesame seed	2.15	1.40	475.65	1.15
Cottonseed	30.30	1.93	1,812.35	1.40
Linseed	2.63	−1.33	63.37	−0.91
Copra	4.67	1.45	1,456.37	1.26

Source: Food and Agriculture Organization of the United Nations, *FAO Production Yearbook* (Rome: FAO, various years).

---

[2]For the purpose of this report, all tons are metric tons.

[3]Palm oil, which comes from the pericarp of the palm tree, is a different product from palm kernel oil, which comes from the kernel.

soybean, palm kernel, sunflower, and linseed production and, to a lesser extent, in groundnut production. In the absence of information on acreage, and given the rapid rate of technological change in palm oil and mounting population pressure on the land during the 1970s and 1980s, it would be plausible to assume that the bulk of the growth in palm oil production is more yield-driven than due to land expansion. Such an assumption is supported by the difference in the relative growth rate of its joint product, palm kernel. In fact, oil palm research has been primarily directed toward higher oil content in the pericarp and smaller kernels.

The expansion of soybean production, on the other hand, derived from both area expansion and productivity improvement, with area expansion accounting for 67 percent of the growth (Kinteh and Badiane 1990). The major dynamism driving area expansion and yield increases in soybean production was the development of high-yielding and early-flowering varieties, which made extension of growing area into northern regions possible. Sunflower seed production was also propelled by rapid biotechnological advances. The crop also experienced significant increases in area of cultivation (Kinteh and Badiane 1990). In addition to the advances in technology, sunflower production was greatly encouraged by strong price, production, and export support systems, particularly in the European Community, as reflected by the escala-tion of European Agricultural Guidance and Guarantee Fund expenditures for oil-seeds during the 1980s.[4]

As one would expect, the change in oilseed production differs significantly across the major producing regions. Virtually all of the expansion in global groundnut production during the 1961-88 period originated from productivity improvements in only three countries—China, India, and the United States (Table 2). Oceania also shows an appreciable increase in area and productivity, but from a relatively low level of production.

The share of Africa in world production fell from 25 percent in the 1960s to 21 percent during the 1982-87 period. Of the overall decline of production in Africa, shrinking cultivated area accounted for about 51 percent, and decline in yields ac-counted for 49 percent. The performance of African producers is in sharp contrast with the performance in other major producing regions. China's production, for instance, grew by an annual rate of 5.4 percent during 1961-88 and reached 26.5 percent of world production in 1982-87. Both area and yield have expanded strongly in China. India raised its production by 1.2 percent a year during 1961-88, reaching a 29.6 percent share of global groundnut production in 1982-87, but virtually the entire increase in produc-tion derived from yield improvement. Even though the United States' average share of 8.5 percent seems low compared with the shares of the other two major producing countries, next to China, it has experienced the highest yield growth rate. Productivity in the United States, however, is one-and-a-half to three-and-a-half times as high as in China and India and nearly four times as high as in Africa.

Despite the deteriorating performance of groundnut production over the period, groundnuts remain the most important source of vegetable oils and fats in Africa, as shown in Table 1. The participation of African oilseed growers in the rapid structural change that has characterized the global vegetable oils and fats sector has been low,

---

[4]European Community expenditures on oilseed programs increased from ECU 0.2 billion in 1976 to nearly ECU 4 billion in 1987 (Andies 1987).

Table 2—Production, area, and yield of groundnuts in shell, by selected regions and countries, 1961-88

Region	Production		Area		Yield	
	1982-87 Average	1982-87 Average Share	1982-87 Average	1961-88 Annual Growth Rate	1982-87 Average	1961-88 Annual Growth Rate
	(million metric tons)	(percent)	(million hectares)	(percent)	(metric tons/ hectare)	(percent)
World	20.08	100.00	18.64	0.13	1.08	1.18
Africa	4.15	20.68	5.79	−0.56	0.72	−0.54
North and Central America	1.90	9.45	0.77	0.41	2.47	2.17
United States	1.71	8.50	0.59	0.27	2.90	2.59
South America	0.70	3.46	0.42	−3.26	1.65	1.41
Brazil	0.27	1.33	0.18	−5.56	1.47	0.75
Asia	13.26	66.05	11.61	0.71	1.14	1.84
India	5.94	29.60	7.15	0.06	0.83	1.10
China	5.32	26.49	2.87	2.47	1.87	2.91
Europe	0.02	0.11	0.01	−0.84	2.11	0.50
Oceania	0.04	0.22	0.03	2.37	1.36	3.47

Source: Food and Agriculture Organization of the United Nations, *FAO Production Yearbook* (Rome: FAO, various years).
Note: Numbers may not add to totals due to rounding.

despite some successes in the development of other competing crops. Soybeans and sunflower seeds, with production levels around 409,000 tons in Africa, are far less important than groundnuts and cottonseed, but their production growth rates are well-above global averages. For cotton, the prospects in the fiber market, the comprehensive (in some cases French-supported) institutional arrangements, and the relatively drought-tolerant character of the crop are likely to give cotton a competitive edge over groundnuts in many African countries. Palm oil production has also performed better than groundnuts. Given the expected rapid rate of adoption of existing technologies in oil palm production, competition from both palm and kernel oils is also likely to intensify, at least in the medium term.

Tables 3 and 4 summarize the evolution of production of groundnuts and other oilseeds in individual AGC countries since 1961. Cottonseed production grew rapidly in Mali and Senegal with average outputs in 1982-87 of 122,000 and 20,500 metric tons, respectively, and impressive annual growth rates of 11 and 18 percent (Table 3). Sesame seed production has taken off in Niger, Nigeria, and Sudan. And, even though levels are still insignificant, palm oil production has grown relatively fast in Senegal. In terms of AGC shares in total African production, sesame seed, palm oil, and palm kernels are the most important crops, with shares of approximately 50 percent.

Groundnut production in Africa averaged about 4 million tons in 1982-87 (Table 4). Since 1961, production has declined steadily at an annual rate of 1.1 percent. Except for Sudan, the relative growth rate of groundnut production in all AGC countries is negative, as it is for Africa as a whole, reflecting a steep decline in cultivated area and to a lesser extent in yields.

The situation in the West Africa subregion is generally worse than the overall African picture. Production in exporting West African countries and Sudan averaged about 2.4 million tons per year during 1982-87, slightly below 60 percent of total

Table 3—Production of major oilseeds and oleaginous fruits other than ground-nuts in AGC countries, 1961-88

Country	Commodity	Production		
		1982-87 Average Share of Africa	1982-87 Average	1961-88 Annual Growth Rate
		(percent)	(1,000 metric tons)	(percent)
The Gambia	Palm kernels	0.30	2.00	0.14
	Cottonseed[a]	0.05	1.12	1.79
Mali	Cottonseed	5.57	122.00	11.16
Niger	Sesame seed[b]	0.04	0.20	7.15
	Cottonseed	0.09	2.03	−3.07
Nigeria	Soybeans	15.38	57.17	−0.14
	Sesame seed	15.77	75.00	1.36
	Coconuts	5.45	98.83	0.73
	Copra	5.64	11.77	1.13
	Palm kernels	50.46	331.50	−0.58
	Palm oil	50.58	736.67	0.70
	Cottonseed	2.64	49.63	−3.16
Senegal	Soybeans[c]	0.05	0.17	. . .
	Coconuts	0.25	4.53	−0.82
	Palm kernels	0.86	5.63	−0.03
	Palm oil	0.41	6.00	2.02
	Cottonseed	0.94	20.50	17.59
Sudan	Castor bean	13.88	5.22	−3.05
	Sunflower seed[d]	2.25	10.17	. . .
	Sesame seed	37.81	179.83	0.23
	Cottonseed	15.37	337.00	−0.15

Source: Food and Agriculture Organization of the United Nations, *FAO Production Yearbook* (Rome: FAO, various years).
Note: AGC is African Groundnut Council.
[a]1977-87
[b]1971-87
[c]1983, 1984
[d]1986, 1987

African production for the period. However, since 1961, the production of West African exporters including Sudan, has been declining at an annual rate of 2 percent.[5] Production of the six non-AGC West African exporters grew by 2.1 percent annually during 1961-88, contrasting sharply with a growth rate of -2.6 percent per year in AGC countries. With the exception of Guinea-Bissau, production expanded in all non-AGC countries, led by Côte d'Ivoire.

In summary, major trends in global and AGC oilseed production include (1) significant shifts in the geographical distribution and commodity-mix of oilseed production; (2) strong productivity gains in groundnut production in North and Central American and Asian countries; (3) a decline in the share of groundnuts in global oilseed production and a fall in the AGC countries' share in African and world groundnut production; and (4) a faster increase in the production of nongroundnut oilseeds than in groundnuts in AGC member countries.

---

[5]Although Sudan is not a West African country, it is a member of AGC, and data for Sudan are often included with the West African countries.

Table 4—Production, area harvested, and yield, in shell, of African groundnut production, selected countries, 1961-88

Country	Production		Area		Yield	
	1982-87 Average	1982-87 Average Share of Africa	1982-87 Average	1961-88 Annual Growth Rate	1982-87 Average	1961-88 Annual Growth Rate
	(1,000 metric tons)	(percent)	(1,000 hectares)	(percent)	(metric tons/ hectare)	(percent)
Total Africa	4,151.71	100.00	5,788.56	−0.56	0.718	−0.54
AGC countries						
The Gambia	112.73	2.72	86.25	−0.51	1.308	−0.28
Mali	86.01	2.07	109.27	−1.70	0.805	0.86
Niger	57.71	1.39	144.31	−4.40	0.409	−2.37
Nigeria	608.00	14.64	698.00	−5.05	0.865	−0.48
Senegal	743.95	17.92	897.72	−0.82	0.854	−0.41
Sudan	394.33	9.50	652.71	3.93	0.614	−1.85
Total AGC countries	2,002.73	48.24	2,588.26	−1.92	0.779	−0.64
West African exporters						
Benin	51.45	1.24	79.12	0.31	0.640	2.58
Guinea	74.45	1.79	129.20	0.83	0.575	−0.47
Guinea-Bissau	28.83	0.69	68.32	−1.86	0.448	−0.59
Burkina Faso	109.52	2.64	177.34	1.98	0.611	0.66
Côte d'Ivoire	101.33	2.44	107.67	3.48	0.940	2.70
Togo	25.76	0.62	51.11	1.63	0.496	0.44
Total non-AGC West African exporters	391.35	9.43	612.76	1.13	0.636	0.97
Total AGC and non-AGC West African exporters	2,394.08	57.67	3,201.02	−1.47	0.751	−0.59

Source: Food and Agriculture Organization of the United Nations, *FAO Production Yearbook* (Rome: FAO, various years).
Note: AGC is African Groundnut Council.

4

TRENDS IN WORLD AND AFRICAN GROUNDNUT COUNCIL OILSEED TRADE

Three major events have shaped the evolution of international trade in oilseeds products since the mid-1970s. These include the emergence of the European Community as a significant oilseed producer, increased productive capacity for palm oil in Indonesia and Malaysia, and expanded soybean plantings in South America. All of these events have increased competition on international markets and changed the commodity mix of traded oilseeds. The structure and evolution of world exports of major oilseed products in Table 5 indicate important shifts in the composition of exports. Soybeans, sunflower seeds, and palm oil have achieved tremendous increases in export volumes in all product categories. In contrast, trade volumes of groundnut seeds and cake fell sharply and nearly stagnated for oil.

At the individual commodity level, there has been a general tendency to export less seed and fruit and more oil and cake. These emerging changes in the product mix of exports have strong implications for the long-term competitiveness of individual exporters. The burden of competition is progressively shifting from production at the farm level to the processing industries. This evolution will inevitably affect the weight of different oilseeds and, of individual exporters on world markets.

The range of groundnut products—in shell or shelled or as oil and cake—broadly reflects different stages of processing. Groundnut oil and its by-product, cake, are the

Table 5—World exports of 10 major oilseeds and oleaginous fruit products, 1961-87

Commodity	1982-87 Average			1961-87 Annual Average Growth Rate		
	Seed	Oil	Cake	Seed	Oil	Cake
	(1,000 metric tons)			(percent)		
Groundnuts	80.37[a]	389.00	631.84	−1.45	−0.05	−3.94
Soybeans	27,206.79	3,594.44	22,012.87	7.92	9.53	12.04
Coconuts	100.16[b]	1,324.46	1,059.06	4.96	5.74	4.51
Palm oil	n.a.	4,889.33	n.a.	n.a.	10.65	n.a.
Palm kernels	118.81	579.32	843.32	−7.80	8.85	6.30
Castor beans	118.21	166.91	n.a.	−1.29	1.18	n.a.
Sunflower seed	1,979.04	1,750.42	1,493.86	10.28	7.09	3.65
Sesame seed	318.73	8.37	59.23	2.53	8.23	0.90
Cottonseed	228.03	361.76	784.43	−3.70	2.26	−1.49
Linseed	676.33	262.93	632.49	0.33	0.24	0.39

Source: Food and Agriculture Organization of the United Nations, *FAO Trade Yearbook* (Rome: FAO, various years).
Note: n.a. indicates that data were not available.
[a]Applies to groundnuts in shell. Shelled groundnuts averaged 762,000 tons and fell at a rate of 2.85 percent annually.
[b]Applies to exports of nuts only. Exports of desiccated nuts and copra amounted to 157,000 and 352,000 tons, respectively, with annual growth rates of 1.25 and 6.67 percent.

14

main groundnut products that are traded; they had a combined export volume of more than 1 million metric tons in 1982-87 (Table 6). Exports of the other two product categories amounted to 761,000 tons for shelled groundnuts and 80,000 tons for groundnuts in shell. The four last columns of Table 6 show that exports of groundnuts in shell and groundnut cake declined for many exporters between 1961 and 1987, probably reflecting the expansion of processing activities and of the livestock sectors in these countries.

The regional export shares vary enormously among regions and products, reflecting, in part, the degree of consumption within the region itself, the extent to which some exporters are dependent on groundnut processing, the type of export market being catered to, and the desire on the part of exporters to increase the value-added of their exports. The evolution of product-specific shares is also affected by domestic agricultural and trade policies in both exporting and importing countries.

North and Central America account for two-fifths of world exports of in-shell groundnuts, and Asia accounts for two-fifths of shelled groundnuts. Despite the decline in production and exports, Africa still represents the largest groundnut oil-exporting region, with a combined share well above one-third of global exports. Groundnut cake exports originate mainly from Asian and African exporters, the former accounting for nearly half of world exports.

Export growth performance during the study period varied significantly across regions. Whereas African exports in all product categories declined by rates ranging from 3 percent for oil to nearly 12 percent for shelled groundnuts, South American exporters achieved significant increases in all their exports except groundnut cake. Shelled groundnut exports from North and Central America and Asia rose about 10 percent a year, in contrast with sharp decreases in Asian exports of groundnuts in shell and cake.

On the import side, Europe was the largest importer in all categories, even though European demand for cake and shelled groundnuts has fallen (Table 7). European imports of groundnuts in shell have risen moderately, while demand for oil has almost stagnated. Although still relatively low, demand for groundnuts in shell has expanded most rapidly in South American markets. Demand for groundnut oil is quite strong in Asia, with the bulk of incremental demand stemming from Japan. Import demand growth rates have, however, been fairly low in almost all regions, except for a few countries such as Italy and to a lesser extent Japan. The slowing of demand in the traditional importing countries of Europe suggests that AGC exports will probably face much tighter competition in the future.

Evolution of the Oilseed Trade in African Countries

Groundnuts and oil palm are by far the most important oilseeds grown and exported by African countries. Soy and sunflower products also play a large role in oilseed imports, and their demand expanded rapidly during the period. This section focuses on these four products.

Although the position of African exporters on international markets is still strong for groundnut oil and palm kernels, it is declining (Table 8). During 1982-87, 36 percent of the nearly half a million tons of world groundnut oil exports originated in Africa, as did virtually all palm kernel exports (Table 6). The trend is falling in all

Table 6— World exports of groundnut products by regions and selected countries, 1961-87

Region/Country	Average Exports, 1982-87								Annual Growth Rate, 1961-87			
	In Shell		Shelled		Oil		Cake		(percent)			
	Quantity	Share	Quantity	Share	Quantity	Share	Quantity	Share	In Shell	Shelled	Oil	Cake
	(1,000 tons)	(percent)	(1,000 tons)	(percent)	(1,000 tons)	(percent)	(1,000 tons)	(percent)				
World	80.37	100.00	761.73	100.00	389.00	100.00	631.84	100.00	-1.45	-2.85	-0.05	-3.94
Africa	5.48	6.82	102.30	13.43	141.71	36.43	221.32	35.03	-11.14	-11.74	-3.14	-3.18
North and Central America	33.06	41.13	227.79	29.90	12.28	3.16	18.10	2.87	8.28	10.34	4.85	15.31[a]
United States	31.51	39.20	22.75	29.87	12.28	3.16	18.10	2.87	23.05	11.32	4.86	2.21[b]
South America	16.21	20.17	97.89	12.85	88.87	22.85	59.31	9.39	23.40[c]	12.73	5.60	-5.91
Brazil	12.00	14.93	1.67	0.22	48.14	12.37	26.99	4.27	-2.73[d]	-6.50	3.69[e]	-6.54
Asia	21.96	27.33	304.37	39.36	78.66	20.22	311.05	49.23	-2.84	8.38	2.02	-4.02
India	0.94[f]	1.17	26.60	3.49	...	...	254.34	40.25	...	-2.17[g]	-14.16[h]	-4.03
China	0.001[i]	0.001	184.43	24.21	64.82	16.66	44.01	6.97	1.21[j]	13.99	7.54	85.25[k]
Europe	3.36	4.19	25.50	3.35	67.11	17.25	22.06	3.49	1.46	3.51	2.88	-6.29
Oceania	0.29	0.37	3.87	0.51	...	...	0.38	0.10	-43.99[l]	26.75	...	11.00[m]

Source: Food and Agriculture Organization of the United Nations, *FAO Trade Yearbook* (Rome: FAO, various years).

Note: All tons are metric tons.

[a]1975-87; [b]1976-87; [c]1969-87; [d]1974-87; [e]1969-87; [f]1982-84; [g]1961-77; [h]1961-79; [i]1986; [j]1963-74; [k]1980-87; [l]1981-87 and growth rate of -28.80 for 1961-79; [m]1979-87.

Table 7—World imports of groundnut products by region and selected countries, 1961-87

Region/Country	Average Imports, 1982-87				Annual Growth Rate, 1961-87			
	In Shell	Shelled	Oil	Cake	In Shell	Shelled	Oil	Cake
	(1,000 metric tons)				(percent)			
Africa	0.60	26.77	21.80	4.40	-10.35	1.98	1.35	-3.65
North and Central America	7.05	71.64	6.40	...	9.56	2.02	-1.74	8.62[a]
South America	9.30	2.37	0.27	...	10.64	7.75	-12.57	...
Asia	20.87	188.13	48.56	67.00	0.85	4.74	1.97	1.29
Japan	...	55.75	0.22	...	...	7.51	2.16[b]	-13.85[c]
Europe	67.23	391.35	308.37	544.57	0.89	-5.31	0.06	-4.46
France	5.99	54.84	153.03	50.02	-0.99	-10.09	0.79	-4.86
Germany	13.20	56.21	22.06	70.89	3.99	-0.97	-2.80	-4.91
Italy	19.50	15.86	35.72	22.10	5.27	-10.02	21.94	12.15
United Kingdom	5.02	90.33	12.17	1.71	0.77	-1.77	-8.39	-24.51
U.S.S.R.	...	56.86	...	8.67[d]	...	3.69	...	2.55[e]
Oceania	0.11	9.66	1.95	...	-4.54	2.83	-8.41	...

Source: Food and Agriculture Organization of the United Nations, *FAO Trade Yearbook* (Rome: FAO, various years).
[a] 1961-73; [b] 1969-87; [c] 1961-79; [d] 1982, 1983, 1985; [e] 1961-75.

product categories except palm kernel cake. AGC shares in African exports of groundnut and oil palm products have been substantial, ranging from 45 to more than 90 percent for groundnut products and from 17 to 70 percent for palm kernel products. The largest AGC exporters are Nigeria (for all palm kernel products), Senegal (for groundnut oil and cake), and Sudan (for nonprocessed groundnuts).

AGC and African exports of nonprocessed oilseeds and fruits have also declined for all products, except during the 1960s and early 1970s for shelled groundnuts in Niger and unshelled groundnuts and palm kernel cake in Senegal. The decline in groundnut oil exports was particularly sharp in Nigeria. The only country with a strong positive export growth rate for that product category was Mali, which started from very low levels. The rapid expansion of palm kernel cake and oil exports by the AGC countries is notable, compared with the 3-4 percent decline for groundnut oil and cake exports.

Except for soybeans, African imports of oilseed products consist mainly of oil (Table 9). In 1982-87, the main vegetable oil imported by African countries was soybean oil, followed by sunflower seed oil and palm oil. At slightly more than 1 percent annually, groundnut oil imports were relatively small and primarily directed to West African countries. Oil imports to African countries, excluding groundnut oil, expanded rapidly throughout the study period, with growth rates ranging from 9 to 18 percent. For the three main West African importers, however, the growth rate for groundnut oil imports exceeded 10 percent.

Nigeria is by far the largest importer of vegetable oils among the three main West African importers, accounting for well over 90 percent of groundnut, soybean, and palm oil imports to the subregion. Moreover, about 40 percent of total African imports of groundnut and palm oil go to that country. While Nigeria still shows a strong preference for groundnut oil, demand has shifted to soybean and palm oil.

Table 8— African and AGC exports of groundnuts and oil palm products, 1961-87

Country/Commodity	Average Exports			Annual Growth Rate of Exports	
	Quantity	Share of African Exports	Data Period	Percent	Data Period
	(1,000 metric tons)	(percent)			
Africa					
Groundnuts					
In shell	5.48	100.00	1982-87	-11.14	1961-87
Shelled	102.30	100.00	1982-87	-11.74	1961-87
Oil	141.71	100.00	1982-87	-3.14	1961-87
Cake	221.32	100.00	1982-87	-3.18	1961-87
Palm					
Kernels	86.53	100.00	1982-87	-8.67	1961-87
Kernel oil	56.20	100.00	1982-87	-0.11	1961-87
Kernel cake	92.67	100.00	1982-87	2.50	1961-87
Palm oil	103.33	100.00	1982-87	-4.94	1961-87
The Gambia					
Groundnuts					
Shelled	25.94	25.35	1982-87	-2.80	1961-87
Oil	7.79	5.50	1982-87	0.42	1961-87
Cake	10.68	4.83	1982-87	2.30	1961-87
Palm					
Kernels	0.10	0.12	1982, 1984, 1985	-6.73	1961-82
Mali					
Groundnuts					
Shelled	1.78	1.74	1982-87	-15.00	1961-87
Oil	1.63	1.15	1982-87	11.85	1966-87
Cake	6.11	2.76	1982-87	7.23	1961-87
Niger					
Groundnuts					
In shell	0.06	0.36	1983, 1984	n.a.	n.a.
Shelled	0.14	0.14	1982-84	0.65	1961-73
Oil	0.02	0.02	1983	-4.50	1961-79
Cake	0.83	0.38	1982-87	-9.93	1961-87
Nigeria					
Groundnuts					
Shelled	1.05	1.03	1982, 1983, 1986	-15.42	1961-74
Oil[a]	n.a.	n.a.	n.a.	-14.76	1961-75
Cake	0.40	0.18	1983	-13.02	1961-77
Palm					
Kernels	56.47	65.26	1982-87	-8.45	1961-87
Kernel oil	17.45	31.05	1982-87	7.06	1962-87
Kernel cake	38.23	41.26	1982-87	11.86	1961-87
Palm oil	2.01	1.95	1982, 1986	-34.90	1961-76
Senegal					
Groundnuts					
In shell	0.06	1.21	1982, 1984	15.01	1966-78
Shelled	7.60	7.42	1982-87	-20.30	1961-87
Oil	111.16	78.44	1982-87	-2.10	1961-87
Cake	134.98	60.99	1982-87	-2.41	1961-87
Palm					
Kernels	1.10	2.27	1985	0.40	1961-71
Kernel oil	0.006	0.01	1983, 1984	n.a.	n.a.
Kernel cake	1.61	1.74	1982-86	16.84	1963-73
Palm oil	0.004	0.004	1982-84	-29.29	1982-84

(continued)

18

Table 8—Continued

Country/Commodity	Average Exports			Annual Growth Rate of Exports	
	Quantity	Share of African Exports	Data Period	Percent	Data Period
	(1,000 metric tons)	(percent)			
Sudan					
Groundnuts					
In shell	7.14	n.a.	1983	−0.42	1961-81
Shelled	24.13	23.59	1982-87	−8.93	1961-87
Oil	7.40	5.22	1982-87	−0.17	1973-87
Cake	56.78	25.65	1982-87	3.40	1961-87
AGC countries					
Groundnuts					
In shell	2.46	44.95	1982-84	−9.33	1961-84
Shelled	60.04	58.69	1982-87	−14.34	1961-87
Oil	127.98	90.31	1982-87	−3.60	1961-87
Cake	209.44	94.63	1982-87	−2.89	1961-87
Palm					
Kernels	56.71	65.53	1982-87	−8.51	1961-87
Kernel oil	17.46	16.89	1982-87	7.06	1962-87
Kernel cake	39.58	70.42	1982-87	11.72	1961-87
Palm oil	1.01	1.09	1982-84	−34.86	1961-76

Source: Food and Agriculture Organization of the United Nations, *FAO Trade Yearbook* (Rome: FAO, various years).
Notse: n.a. indicates that data were not available. AGC is African Groundnut Council.
[a]Nigeria did not export during the second half of the 1980s, the period for which export shares are computed.

In summary, the following major changes have occurred in global and AGC oilseed trade:

(1) increased competition on world groundnut markets by South American and Asian exporters;

(2) rapid expansion of soybean, sunflower, and palm oil production in the EC and Asian countries, mainly Indonesia and Malaysia, which has put international oil and cake markets under increased pressure;

(3) shift of demand for oilseeds in the traditionally largest groundnut-importing European countries toward competing products;

(4) strong expansion of vegetable oil demand in African countries, associated with a rapid increase in demand for groundnut oil in West Africa;

(5) an emerging shift of regional demand in West Africa toward palm and soybean oil;

(6) decline in exports of unprocessed groundnuts in all AGC countries; increases in groundnut oil exports from a low level in Mali and to a lesser extent in The Gambia; and

(7) increases in palm kernel products in the exports of AGC member countries.

Table 9—Imports of major oilseeds and oleaginous fruit products by selected African countries

Country/Commodity	Average Imports			Annual Growth Rate of Imports	
	Quantity	Share of African Imports	Data Period	Percent	Data Period
	(1,000 metric tons)	(percent)			
Africa					
Soybeans					
Beans	55.30	100.00	1982-87	13.56	1961-87
Oil	388.86	100.00	1982-87	8.64	1961-87
Cake	692.50	100.00	1982-87	37.22	1961-87
Groundnuts					
In shell	0.60	100.00	1982-87	-10.35	1961-87
Shelled	26.77	100.00	1982-87	1.98	1961-87
Oil	21.80	100.00	1982-87	1.35	1961-87
Cake	4.40	100.00	1982-87	-3.65	1961-87
Palm					
Kernels	0.02	100.00	1982, 1983, 1986, 1987	-18.48	1961-75
Kernel oil	42.58	100.00	1982-87	12.56	1961-87
Palm oil	351.62	100.00	1982-87	18.15	1961-87
Sunflower					
Seeds	22.45	100.00	1982-87	21.35	1961-87
Oil	355.42	100.00	1982-87	12.17	1961-87
Cake	4.50	100.00	1982, 1985-87	10.24 / -6.82	1978-82 / 1968-72
Ghana					
Soybean oil	3.27	0.84	1982-87	10.68	1961-87
Groundnut oil	0.10	0.47	1982, 1983	-3.13	1961-83
Palm oil	4.30	1.22	1982-87	13.61	1961-87
Côte d'Ivoire					
Soybean oil	0.58	0.15	1982-87	-4.10	1971-87
Groundnuts					
Oil	0.12	0.55	1982-87	-13.97	1964-87
Cake	0.42	9.59	1974-83	-19.74	1974-83
Nigeria					
Soybeans					
Oil	23.03	5.92	1982-87	-2.05	1975-87
Cake	17.50	2.53	1982-87	-0.32	1978-87
Groundnut oil	9.10	41.72	1982-87	35.49	1963-87
Palm oil	136.83	38.92	1982-87	50.89	1977-87
Total of three countries					
Soybeans					
Oil	26.89	6.92	1982-87	18.84	1961-87
Cake	17.50	2.52	1982-87	0.32	1978-87
Groundnuts					
Oil	9.25	42.42	1982-87	10.21	1961-87
Cake	0.42	9.59	1982, 1983 1985	13.88 / -19.74	1966-72 / 1974-83
Palm oil	141.13	40.14	1982-87	35.32	1961-87

Source: Food and Agriculture Organization of the United Nations, *FAO Trade Yearbook* (Rome: FAO, various years).

5

EXTERNAL DEMAND CONSTRAINT, DOMESTIC POLICIES, AND EXPORT PERFORMANCE OF AFRICAN GROUNDNUT COUNCIL COUNTRIES

Production and Trade Performance

African countries, mainly AGC members, produce as much as 20 percent of the world groundnut supply and are still major players in international markets, despite the significant changes that occurred in the world oilseed economy between the 1960s and the late 1980s (Table 10). The picture is much less impressive, however, if one looks at the long-term changes in the groundnut economy of the AGC countries.

As shown earlier in Table 8, groundnut oil exports from member countries fell on average by nearly 4 percent annually between the early 1960s and the late 1980s. The growth rates of the two largest exporters, Nigeria and Senegal, were −15 and −2 percent, respectively. Nigeria turned from being the largest exporter among the AGC countries, with a share of 26 percent in world exports, to a net importer in the 1980s. At the same time, competitors from Asia and South America were able to boost their exports by nearly 400 percent, raising their combined share from slightly over 10 percent in 1961-65 to 50 percent of world exports of groundnut products in 1986-88. As a result, the AGC's export share fell from 62 percent to 20 percent.

Table 10— Groundnut oil trade and production performance, AGC countries compared with selected regions and the world, 1961-65 and 1986-88

Country	Production Growth Rate	Yield Growth Rate	Share of World			
			Production		Exports[a]	
			1961-65	1986-88	1961-65	1986-88
	(percent)			(percent)		
AGC countries	−2.61	−0.32	23	10	62	20
The Gambia	−0.71	−0.19	1	1	2	2
Mali	−2.21	0.05	1	0	2	1
Niger	−7.03	−2.56	1	0	4	0
Nigeria	−5.47	0.36	12	3	26	0
Senegal	−1.14	−0.20	6	4	23	14
Sudan	2.02	−1.71	2	2	5	3
South America	−1.85	1.43	7	3	5	18
Asia	2.46	1.78	51	67	8	32
World	1.24	1.19	100	100	100	100

Source: Food and Agriculture Organization of the United Nations, *FAO Production Yearbook* (Rome: FAO, various years); Food and Agriculture Organization of the United Nations, *FAO Trade Yearbook* (Rome: FAO, various years).

Note: AGC is African Groundnut Council.

[a]Groundnut exports are in oil equivalents.

The fall in groundnut exports is paralleled by a marked decrease in groundnut production in all AGC countries, except Sudan, as can be seen in the first two columns of Table 10. The decline was substantial in Niger (−7 percent) and Nigeria (−5 percent). Aggregate production in the AGC countries as a whole decreased at an annual rate of 2.6 percent, compared with a growth rate of 1.2 percent for the world and 2.5 percent for Asia. The poor performance of the AGC countries is also reflected in groundnut yields, which increased in both Asia and South America, but fell in AGC countries by 0.3 percent a year compared with a world average growth rate of 1.2 percent. With the exception of Sudan, AGC countries have also experienced a rapid decrease in cultivated area, indicating a deterioration of incentives to farmers to keep investment in the groundnut sector. Consequently, the share of AGC countries in world groundnut production fell from 24 percent in 1961-65 to only 10 percent in 1982-87. The sharpest decrease in production occurred in Nigeria, whose share fell from 12 to 3 percent.

These dramatic changes in AGC exports and production took place despite the relative stability of world demand for groundnut products during the period. The overall decline in world demand for groundnut oil from 1961 to 1987 only averaged 0.05 percent a year (see Table 6). This, together with the impressive expansion of Asian and South American trade shares, sharply contradicts the argument of external demand constraint.

Therefore, an attempt is made next to isolate the contributions to AGC's trade decline by comparing the effects of changes in export market conditions with the effects of internal factors, such as macroeconomic policies and trade strategies in member countries. First, a simple model decomposing the changes in the real value of groundnut exports, the income terms of trade shown in Figures 1-4, is calculated. The decomposition highlights the relative contribution of changes in the external terms of trade and of changes in domestic supply for export markets. Then a model estimating the impact of policies on prices and incentives within and outside the groundnut sector is used to show the role of domestic policies in the decline of groundnut production and trade in AGC countries.

For each member country and the group of AGC countries, the index of the real value of groundnut exports in the base period $t0$ and any subsequent year tn can be expressed as[6]

$$G_{x,t0} = P_{r,t0} \cdot Q_{x,t0} \,, \tag{1}$$

$$G_{x,tn} = e^{\alpha t} P_{r,t0} \cdot e^{\beta t} Q_{x,t0} \,, \tag{2}$$

where $G_{x,t0}$ is the index of the value of groundnut exports deflated by the prices of country imports; $P_{r,t0}$ stands for the terms of trade, defined as the ratio in t0 between the unit values of groundnut exports and those of country imports; $Q_{x,t0}$, is the volume index of country groundnut exports in the base year.; and α and β are the constant exponential growth rates of the terms of trade and export volume indices. Defining the rate at which the value of country groundnut exports grew as γ, then equation (2) can be rewritten as

$$G_{x,tn} = e^{\gamma t} P_{r,t0} \cdot Q_{x,t0} \,. \tag{3}$$

---

[6]See Svedberg 1991.

22

This gives the identity $\gamma = \alpha + \beta$. The ratios α/γ and β/γ can be used to measure the contributions of changes in country terms of trade and the supply of exports.

The same approach can be used in analyzing AGC countries' trade performance relative to other groundnut exporters and other oilseed exporters to find out whether AGC countries or groundnut markets have specific characteristics that led to the observed loss of performance. Using equations (1) and (2), one can write the real value of AGC countries' exports relative to that of other groundnut or oilseed exporters:

$$G_{x,t0}^a/G_{x,t0}^c = P_{r,t0}^a/P_{r,t0}^c \cdot Q_{x,t0}^a/Q_{x,t0}^c \,, \tag{4}$$

where $G_{x,t0}^a$ is the value of AGC groundnut exports and $G_{x,t0}^c$ is the value of groundnut or oilseed exports by the comparator countries—the other exporters; $P_{r,t0}^a/P_{r,t0}^c$ is the ratio of AGC terms of trade to that of the other exporters; and $Q_{x,t0}^a/Q_{x,t0}^c$ is the ratio of supplied export quantities between the two groups of countries. Following equation (3) and defining f, h, and g as the growth rates of the ratios between the P, Q, and G variables in equation (4), the latter can be written for any period n as

$$G_{x,tn}^a/G_{x,tn}^c = e^{ft}(P_{r,t0}^a/P_{r,t0}^c) \cdot e^{ht}(Q_{x,t0}^a/Q_{x,t0}^c). \tag{5}$$

Here again $g = f + h$ is an identity, and the ratios f/g and h/g can be used to measure the relative contributions of changes in the terms of trade that are exogenous to AGC and non-AGC countries alike and of changes on the export supply side that are more domestically driven.

Before discussing the results of these computations, it may be useful to briefly review some relevant changes on international groundnut and oilseed markets, as presented in Figures 5-8. Figures 5 and 6 compare the trends in world groundnut prices to those of other oilseeds. First, after the early 1970s prices on world groundnut and other oilseed markets increased to approximately twice their level of the 1960s. Second, prices were higher for groundnuts than for other oilseeds. Third, the differences between groundnut and other oilseed prices increased over time. Turning to export volumes, aggregate demand for oilseeds increased significantly during the study period (Figure 7). While world exports of groundnuts expanded less dramatically, quantities supplied by AGC countries fell sharply after the first half of the 1970s (Figure 8).

The observed trends in world prices and in global demand for groundnuts and other oilseeds do not seem to have constituted a serious constraint to export expansion in AGC countries. On the other hand, the results of the decomposition of the real value of exports (Table 11) indicate that changes in the export volumes of AGC countries were a significant factor. The first column of the table shows growth rates for the real value of exports of the AGC countries as a whole and for selected countries. Ratios of competing exporters of groundnuts and other oilseeds range from –2 to –13 percent. For individual countries as well as for the AGC as a whole, changes in export quantities contributed 80 percent and more to the decline (column 2). The contribution is higher than 100 percent in cases where negative changes in supplied quantities have overcompensated for positive changes in the terms of trade, causing a fall in real and relative export values. These correspond to the 7 cases out of the 11 in the last column that have negative coefficients. In all of these cases, the value and volume of AGC countries' exports fell both absolutely and relatively, despite increases in the absolute and relative terms of trade.

Figure 5—World prices of groundnuts compared with soybeans and palm, 1962-87

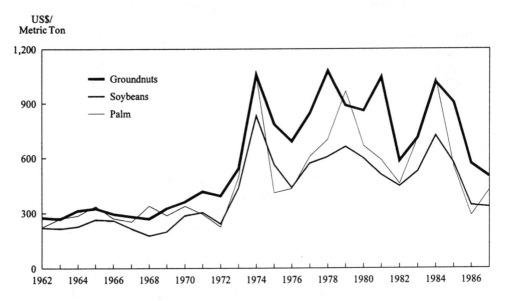

US$/
Metric Ton

Source: Food and Agriculture Organization of the United Nations, *FAO Trade Yearbook* (Rome: FAO, various years); World Bank, *Commodity Trade and Price Trends* (Washington, D.C.: World Bank, various years).

Figure 6—World prices of groundnuts compared with sunflower seed and rapeseed, 1962-87

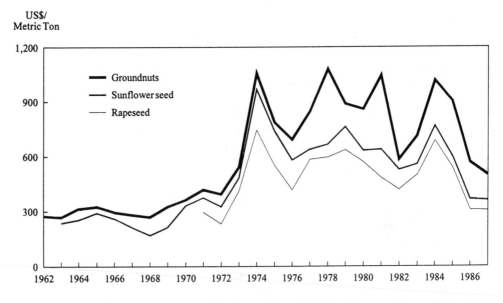

US$/
Metric Ton

Source: Food and Agriculture Organization of the United Nations, *FAO Trade Yearbook* (Rome: FAO, various years); World Bank, *Commodity Trade and Price Trends* (Washington, D.C.: World Bank, various years).

Figure 7—World exports of oilseeds, 1961-87

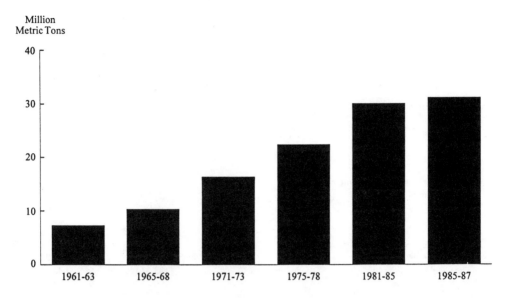

Million
Metric Tons

Source: Food and Agriculture Organization of the United Nations, *FAO Trade Yearbook* (Rome: FAO, various years).

Figure 8—Non-AGC and AGC exports of groundnuts, 1961-87

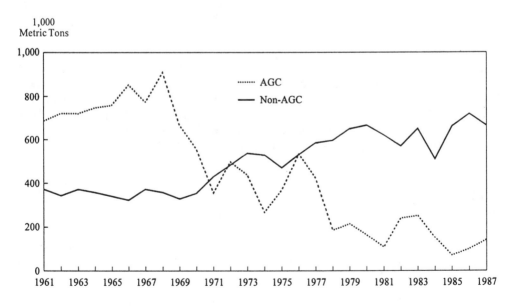

1,000
Metric Tons

Source: Food and Agriculture Organization of the United Nations, *FAO Trade Yearbook* (Rome: FAO, various years).
Note: AGC is African Grounndut Council.

Table 11— Decomposition of annual changes in the real value of groundnut exports, 1961-87

Change	Change in Export Value	Contribution of Volume Changes[a]	Contribution of Country Terms of Trade Changes[a]
		(percent)	
Change in the real value of exports			
AGC countries	−8.12	83.31	16.69
The Gambia	−1.81	116.46	−16.46
Senegal	−5.59	78.56	21.44
Sudan	−4.12	112.68	−12.68
Change in the value of exports			
Relative to other groundnut exporters			
AGC countries	−10.40	77.76	22.24
The Gambia	n.s.	n.s.	n.s.
Senegal	−4.58	79.18	20.82
Sudan	−2.25	148.19	−48.19
Relative to other oilseed exporters			
AGC countries	−13.29	103.38	−3.38
The Gambia	−6.50	133.72	−33.72
Senegal	−10.28	106.75	−6.75
Sudan	−8.92	127.00	−27.00

Source: Food and Agriculture Organization of the United Nations, *FAO Trade Yearbook* (Rome: FAO, various years).
Notes: n.s. is "not significant." All trend coefficients are significant, except for The Gambia. AGC is African Groundnut Council.
[a]Calculated as annual changes in export volume and terms of trade divided by corresponding annual changes in real value of exports in the first column.

An increase in the relative terms of trade and therefore a negative contribution to the decline of export values was observed mainly in the comparison with other oilseeds, due to the faster increase in world groundnut prices shown in Figures 5 and 6. And, even though the real value of Senegal's exports fell faster than those of The Gambia and Sudan, the quantity contribution was more pronounced in the latter two countries. The results obtained so far indicate that supply-side factors must have played a predominant role in the decline of the AGC countries' export performance.

Given the relationships that exist between domestic economic policies and the performance of export sectors such as the groundnut sector, the possible role of these domestic policies in the observed decline in the AGC's trade is investigated next.

The Role of Domestic Policies in the Decline of the Groundnut Industry

This report focuses exclusively on the role of price factors in the decline of the groundnut economy in AGC countries. An earlier study mandated by the AGC, which focused on nonprice factors (UNECA/FAO 1985), found that factors related to the production of and access to improved seeds and to the availability and distribution of fertilizer have played significant roles in the decline of groundnut production and trade in AGC countries (see Appendix 1). By focusing only on price factors and their relationship to domestic policies, the present study seeks to complement the previous study by highlighting the critical role of the incentive environment.

The agricultural sector, and particularly its tradable component, is known to be highly sensitive to the bias that characterizes sector and economy-wide policies in many developing countries.[7] Policies that are expected to affect the AGC countries' performance in groundnut production and trade are groundnut marketing and pricing policies, protection of import-competing sectors, and foreign exchange control and other restrictions to trade that cause the real exchange rate to appreciate. In the following sections, the extent to which policies in AGC countries have played a role in the decline of the groundnut industry is investigated.

First, the impact of country policies on incentives in the groundnut sector is estimated by comparing the actual levels of real groundnut prices to the ones that would prevail in the absence of these policies. Then, estimates of the effects on groundnut production and exports of individual countries and the AGC as a whole are derived.[8] The analysis of the impact on incentives in the groundnut sector distinguishes between direct and indirect effects of domestic policies. The direct effects correspond to effects emanating directly from measures in the groundnut sector, such as pricing and marketing. The indirect effects result from the deviation of country exchange rates from their equilibrium caused by overall trade and foreign exchange policies. The sum of the two determines the ultimate effect of domestic policies on groundnut producer incentives.

In Senegal, the years following the country's independence in 1960 were marked by a rapid expansion of public involvement in pricing and marketing of groundnuts. By 1968, public intervention in groundnut marketing had reached a level where private trader participation was proscribed to guarantee complete state monopoly. In 1961, private traders handled about 80 percent of the marketed surplus, with the remaining 20 percent handled by about 700 farmer cooperatives, which were state-sponsored and linked to the parastatals. By 1967/68, 70 percent of the marketed groundnut output was handled through farmer cooperatives, the number of which had nearly doubled to reach 1,300 (Jammeh 1987). For the next 20 years, groundnut procurement and export as well as domestic sales of groundnut products remained under the control of marketing parastatals and the national oil milling company, Société Nationale de Commercialisation des Oleagineux du Sénégal (SONACOS). By 1992, SONACOS had a market share of 90 percent. As a result of the liberalization programs of the second half of the 1980s, public involvement has decreased, but SONACOS, through regional oil millers, still organizes the procurement of groundnuts through a network of private contract traders. Through arrangements linking it to licensed private traders, SONACOS still operates a form of price fixing (Gaye 1991).

The increased involvement of public institutions in groundnut marketing was accompanied by a severe depression of producer prices. Until 1985/86, a main feature of groundnut pricing policies in Senegal was the withholding of a certain percentage of the officially fixed groundnut prices for input loans and quality insurance. Besides payments for cooperatives' credit risk insurance, the withholdings usually covered

---

[7]See the comprehensive comparative study by Krueger, Schiff, and Valdés (1992).

[8]The analysis is based on the approach taken by Krueger, Schiff, and Valdés (1988).

marketing board losses (above a determined limit) and grain quality losses. Contrary to officially stated policy, the withholdings were not repaid to farmers in years when marketing board losses were within permissible limits and the quality of grains satisfactory, but disappeared into an investment fund owned by the national cooperative association (Jammeh 1987). Moreover, calculations by Jammeh (1987) and SOFRECO (1988a) indicate that marketing board margins during this period averaged 45-50 percent of producer prices.

Similarly, groundnut marketing in The Gambia was characterized by heavy involvement on the part of the state. During the period covered by the study, The Gambia Produce and Marketing Board (GPMB) had exclusive rights to purchase and export groundnuts. As in Senegal, procurement is organized through the Gambian Cooperative Union (GCU) and a network of licensed private traders, on the basis of arrangements that ensure control by the marketing board. Consequently, GPMB's share in marketed groundnuts rose from 40 percent in the mid-1970s to 80 percent of the crop a decade later (Jones 1986). In both Senegal and The Gambia, marketing boards strongly influenced the determination of official prices. In order to enforce official prices and facilitate procurement at these prices, trading in both cases was not allowed outside of an officially specified marketing season (Jammeh 1987; Kristjansen et al. 1990). Recent attempts at liberalization have increased private participation in the marketing of groundnuts. GCU's share is still in the 50 percent range (Kristjansen et al. 1990). In both countries, marketing and pricing policies have had strong negative effects on producer incentives, as will be shown later.

Groundnut pricing and marketing policies in Sudan are very different from those observed in The Gambia and Senegal. Despite its restriction to a small group of exporters during most of the 1970s, groundnut marketing in Sudan is carried out exclusively by private traders and organized through a system of rural and urban auction markets. Also, pricing policies in Sudan have been more favorable to groundnut producers. They include a combination of producer price support, exchange rate subsidies, preferential export taxes, and local government and marketing taxes (El Bashir and Idris 1983). The net effect of these policies on producer prices was positive over most of the study period.

However, it is possible for indirect effects emanating from policies outside of the groundnut sector to overcompensate for the positive effect. As explained earlier, the groundnut sector is also affected indirectly by restrictive trade policies that raise prices in other domestic sectors relative to groundnut prices. Protection and foreign exchange controls also hurt the exporting groundnut sector through their effect on the real exchange rate. As in many other developing countries, protection-based import substitution has received marked attention in the development strategies of AGC countries. The nature of trade regimes in the study countries during most of the period covered by this report is treated elsewhere (see Oyejide 1993; Gulhati, Bose, and Atukorala 1985; Elbadawi 1988; and World Bank 1987). This report focuses instead on analyzing the impact of the removal of country trade restrictions on incentives in the groundnut sector.

As stated earlier, the groundnut sector is vulnerable to general imbalances in economy-wide policies that cause country exchange rates to appreciate. Such imbalances are typically reflected in a sustained deterioration of country trade balances. With the exception of a few years for Sudan, data for The Gambia, Senegal, and Sudan show rapid and sustained increases of the deficit in their current accounts.

(Much of the data on which this report is based is contained in a supplement available on request from the International Food Policy Research Institute.) The resulting effects of both trade restrictions and overall economic policies on the groundnut sector can be estimated by calculating the real exchange rates that would prevail if the underlying policy imbalances were corrected to bring country current accounts to sustainable levels.

The next step of the analysis is to show how these direct and indirect effects have affected the groundnut production and export of individual AGC countries. The analysis of the output and export effects is carried out based on estimates of the response of output and exports to changes in relative prices of groundnuts and the size of the price changes that an elimination of the price disincentives discussed earlier would bring about.

The Evolution of Incentives in the Groundnut Sector

Before estimating the effects of policies on groundnut incentives, changes in real groundnut prices for each country are analyzed. Following Quiroz and Valdés (1993), the changes in real groundnut prices are decomposed into (1) changes in the international prices, (2) changes in the nominal rate of protection, (3) changes in domestic transfer costs, and (4) changes in the real exchange rate. The first component shows the contribution of trends in export prices and the second and third components the contribution of policies on domestic marketing and groundnut exports (subsidy/tax) to changes in domestic groundnut prices. The fourth component represents the contribution of changes in the real exchange rates of the countries.

Omitting the subscript t for the time trend, actual real groundnut prices, p_{GN}^a, in each year can be expressed as

$$p_{GN}^a = P_{GN}^a / P_{NA}^a. \tag{6}$$

For each time period, P_{GN}^a denotes the nominal groundnut producer price and P_{NA}^a the general price level in the nonagricultural sector. P_{GN}^a can be defined as a function of the export price P_{GN}^w and the nominal exchange rate E^a:

$$P_{GN}^a = M_{GN}^a T_{GN}^w E^a P_{GN}^w, \tag{7}$$

where

$$M_{GN}^a = (1 + t_{GN}^a), \tag{8}$$

and

$$T_{GN}^w = (1 + t_{GN}^w). \tag{9}$$

t_{GN}^a stands for the ratio of transfer costs to the border equivalent of groundnut producer prices and t_{GN}^w represents the percentage difference between the border-equivalent producer price and the actual export price, P_{GN}^w, that is, the nominal protection rate.

Expressing the general level of nonagricultural foreign prices by P_{NA}^w, expanding equation (6), and using equation (7), one can rewrite the actual real groundnut price as

$$p_{GN}^a = M_{GN}^a T_{GN}^w E^r p_{GN}^w, \tag{10}$$

E^r represents the real exchange rate and is defined as

$$E^r = E^a (P_{NA}^w / P_{NA}^a).\qquad(11)$$

p_{GN}^w is defined similarly to equation (6) as

$$p_{GN}^w = P_{GN}^w / P_{NA}^w.\qquad(12)$$

Given a base period $t0$, the real price of groundnuts in any subsequent period tn can be expressed as

$$p_{GN,tn}^a = (e^{kt} \cdot M_{GN,t0}^a)\,(e^{lt} \cdot T_{GN,t0}^w)\,(e^{ut} \cdot E_{t0}^r)\,(e^{vt} \cdot p_{GN,t0}^w).\qquad(13)$$

The growth rates in equation (13), k, l, u, and v, can be used as indicators of the role of the individual variables on the left-hand side of the equation in the evolution of incentives in the groundnut sector.

Because government intervention in the pricing and distribution of groundnuts has been quite extensive during most of the study period, particularly in The Gambia and Senegal, and because policy-related factors have been a major determinant of the cost of transferring groundnuts to the ports of export, the coefficients k and l are used to reflect the influence of domestic marketing and protection policies on changes in real groundnut prices. Given the definition of M_{GN}^a and T_{GN}^w in equations (8) and (9), k and l will take on negative values with absolute increases in the (negative) transfer cost, t_{GN}^a, and protection rates, t_{GN}^w. Similarly, u and v show the influence of changes in the real exchange rates and the level of groundnut export prices, respectively.

The results of the decomposition of real producer prices for The Gambia, Senegal, and Sudan are presented in Table 12. After a period of rising real producer prices in the 1960s in The Gambia and Senegal, groundnut prices fell rapidly in all three countries during most of the 1970s. The trend reversed in the 1980s, with stable prices in Senegal, strong increases in The Gambia, and moderate increases in Sudan.

During the first two periods, rapid increases in transfer costs had a strong negative impact on real producer prices in The Gambia. Transfer costs, combined with a continuous increase in the rate of taxation of groundnut exports, as portrayed by the decline in the rate of nominal protection in the third column, would have almost entirely eliminated the gains by Gambian producers from the average annual 4 percent rise in export prices in the 1960s. However, Gambian producer prices increased in the 1980s, if only at half the rate of export prices, mainly due to the gradual depreciation of the real exchange rate (column 4). Further increases in the costs of transfer and a rapid decline in export prices translated into a rapid decline of real producer prices during the 1970s, despite the downward trend in the rate of export taxation and the continued depreciation of the Gambian currency, the dalasi (D).[9] In contrast, the accelerated depreciation of the real exchange rate during the following decade more than compensated for the strong increase in the rate of export taxation and the continuous decline in export prices to boost real groundnut prices.

---

[9] In June 1987, US$1.00=D7.50.

Table 12—Decomposition of the changes in real groundnut producer prices, 1968-88

	Changes in					Export Quantity Growth Rate	Output Quantity Growth Rate
Country	Real Producer Price, P^a_{GN}	Transfer Costs, M^a_{GN}	Exort Taxes, T^w_{GN}	Real Exchange Rate, E^r	Export Price, P^w_{GN}		
	(percent)						
The Gambia							
1966-72	2.18	-2.79	-0.73	1.62	4.06	-4.08	-2.31
1973-80	-2.48	-0.99	0.97	0.65	-3.12	-9.34	-10.27
1981-88	5.24	1.21	-3.27	11.13	-3.84	1.39	-2.74
Senegal							
1966-72	3.16	-3.24	7.78	2.78	-4.15	-10.36	-5.32
1973-80	-3.53	-1.47	1.23	0.49	-3.78	-3.88	-6.23
1981-88	0.43	3.30	11.06	-3.23	-10.70	11.00	-0.22
Sudan							
1966-72	-0.94	-2.87	3.47	-2.13	0.58	2.10	7.92
1973-80	-1.52	-2.23	2.99	6.70	-8.97	-11.97	1.95
1981-88	1.39	3.01	15.22	-2.65	-14.19	-20.81	-5.46

Sources: Growth rates for P^a_{GN} were computed from column (3) of Tables 1-3 in the data supplement to this report, available on request from the International Food Policy Research Institute. Computations for M^a_{GN} are based on columns (1) and (4) and T^w_{GN} are based on columns (5) and (7) of those tables. Growth rates for E^r and P^w_{GN} are computed from columns (6) and (7). The latter is obtained by deflating country groundnut export unit values by the world manufacturing unit value index from IMF (International Monetary Fund), *Financial Statistics* (Washington, D.C.: IMF, various years). M^a_{GN} and T^w_{GN} are defined as in equations (8) and (9), where t^a_{GN} and t^w_{GN} are negatively signed. Therefore, negative numbers in the corresponding columns reflect increases in transfer costs or the rate of taxation.

In Senegal, export prices fell sharply during the 1960s, following the elimination of the French export price support system (Jammeh 1987, 157). The drop in export prices, coupled with a significant increase in the costs of transfer, would have led to a severe depression of groundnut prices had their effects not been entirely absorbed through a substantial reduction in the level of export taxation, which, along with the depreciation of the real exchange rate, helped real producer prices in Senegal achieve the fastest growth rate of the three countries. With much smaller changes in the transfer costs, export taxation, and the real exchange rate, most of the decline in export prices during the 1970s was transmitted directly to domestic groundnut producers. In the 1980s, strong cuts in transfer costs and the rate of export taxation helped absorb the combined effects of a rapid appreciation of the real exchange rate and a free fall in export prices to stabilize real producer prices.

In Sudan, rising transfer costs and appreciation of the real exchange rate surpassed the reduction in export taxation, resulting in falling real groundnut prices during the 1960s, despite a continuous increase in the export price. Groundnut prices continued their decline through the 1970s, aided by the sharp decrease in export prices and further increases in transfer costs. However, the rapid depreciation of the real exchange rate and continued reduction of the level of taxation strongly contributed to keeping the level of decline of domestic prices far below that of export prices. For the first time during the study period, real groundnut prices started to rise in the 1980s, despite the significant drop in export prices and the appreciating exchange rate, mainly as a result of a rapid decrease in the taxation of exports and, to a lesser extent, of transfer costs.

Comparing the experiences in the three countries, it can be seen that Senegal and Sudan adjusted to changes in export prices primarily through reductions in the rate of nominal protection or taxation. In The Gambia, on the other hand, continuous depreciation of the exchange rate, especially in the last period, protected domestic producers from negative changes in export prices, while the level of taxation rose during most of the period. In all three countries, rising transfer costs had a strong negative impact on incentives in the groundnut sector throughout the first two decades.

The evolution of the volume of groundnut exports and output by the three countries over the three periods is recorded in the sixth and seventh columns of Table 12. In Sudan, output and exports expanded rapidly with stagnating export and producer prices during the first period. After a modest increase during the 1970s, output decreased sharply in the 1980s. Exports, however, decreased rapidly during both the 1970s and 1980s. Two facts may help explain the differences in the evolution of price incentives and output quantities in Sudan. A significant share of output is grown in the irrigated areas of Gezira. However, production decisions in the irrigated areas are determined primarily by tenancy size and government regulations on crop mix rather than by relative prices (El Bashir and Idris 1983).

Whereas Sudan's decline in exports in the 1980s can be tied to the decline in output in the same period, the fall in exports during the preceding period is partly explained by the second fact. According to El Bashir and Idris (1983), domestic use of groundnuts increased considerably during the 1970s at the expense of exports, due to a sharp drop in cottonseed production and a consequent increase in demand for groundnuts by domestic vegetable oil processors. These occurrences mask the relationships between changes in prices and groundnut output and export quantities—relationships that become more visible in the other two countries, especially in the last two periods.

During the first period, the decline in output and exported quantities in Senegal and The Gambia contrasted sharply with the increases in real groundnut prices. This is a reflection of the severe drought that struck large areas of the region and depressed output for most of the years during that period. The impact of the drop in output was large enough to lead to falling export quantities in The Gambia, despite increasing producer and export prices. Besides the fall in output, exports in Senegal were probably affected by changes on the export front. As mentioned earlier, the fall in Senegalese export prices recorded during that period mainly reflect the elimination of the export price support arrangement with France. With the elimination of the support system, the French companies that were handling the totality of Senegal's groundnut exports began to shift to other sources of supply and away from groundnuts (Jammeh 1987).

During the 1970s, falling export prices were met in both The Gambia and Senegal with modest changes in exchange rates, but continued export taxation and even increases in domestic transfer costs resulted in rapidly decreasing producer prices. The continuation of adverse domestic policies during the period of falling international prices must have significantly contributed to the rapid deterioration of groundnut production and trade performance in the two countries. The implications of the failure to adjust domestic policies to changes in the international trading environment can be seen by contrasting the developments in the 1970s with those of the 1980s. During the latter period, the fall in export prices was faster than in the 1970s, particularly in Senegal. However, by adjusting domestic policies, both countries succeeded in reversing the fall in producer prices. In Senegal, these developments must have played a significant role in slowing down the decline in output and even stopping it.

Export quantities also increased, especially in Senegal. The rapid growth of exports in Senegal probably reflects in part the smuggling of Gambian groundnuts into that country. According to Puetz and von Braun (1990a; 1990b), more than 50 percent of marketed Gambian groundnuts in border areas were smuggled into Senegal. The sale of Gambian groundnuts to Senegal was prompted by the pricing reforms undertaken in the two countries and the strong devaluation of the Gambian currency. During most of the 1980s, Gambian prices were 20-30 percent below prices prevailing in Senegal. Furthermore, due to budgetary difficulties, the Gambian marketing board decided not to purchase large portions of the domestic output, which also contributed to the unrecorded export of groundnuts to Senegal.

It is interesting to note the difference in adjustment strategies. Senegal, being a member of the West African Monetary Union, was unable to use the exchange rate to adjust. In fact, the real exchange rate in that country appreciated rapidly during the considered period. Thus, whereas Senegal adjusted mainly through the budget by cutting export taxes and to a lesser extent reducing distribution costs, The Gambia worked through the exchange rate, expanding its tax revenues from groundnut exports at the same time.

In order to show the impact of these changes on the groundnut sector, it is necessary to examine their impact on the absolute level of relative prices. The impact of domestic policies on absolute prices of groundnuts reveals how AGC exporters fared in comparison with competing foreign exporters, assuming these competitors were not discriminated against and therefore received their asking price. In the next section, the effects on the output and export levels by countries are estimated.

Effects of Domestic Policies on Incentives in the Groundnut Sector

Unlike sectoral policies that affect only prices in the groundnut sector, trade restrictions and other economic policies affect the remaining sectors of the economy as well. Accordingly, the impact of policies on groundnut incentives is measured by the differences between actual relative prices of groundnuts and the relative prices that would obtain in a situation without direct and implicit taxation arising from domestic policies. For this purpose, equivalent border prices are used as reference prices where there is no intervention. The impact of the changes in relative prices on groundnut output and exports is estimated using estimates of supply elasticities for groundnuts.

To begin, groundnut supplies in individual countries are expressed as functions of relative prices:[10]

$$Q_{GN}^a = \Sigma_i \alpha_i X_i^a + \mu_{GN}.$$ (14)

---

[10]Because of the unavailability of data on input use, only output prices are used in supply equations. The effect on the estimates is likely to be negligible, since the expected low levels of intermediate input use render the impact of policy changes at this level relatively insignificant, compared with the effects on output prices.

As in equation (6), the index t for the time period is dropped, and the superscript a refers to actual values of the corresponding variables. Q^a_{GN} is the actual level of groundnut output in each country; α_i represents the elasticities of supply with respect to X^a_i the exogenous variables; and μ_{GN} is the estimation residual. In addition to the variable for the constant, X^a_1 and the one for the lagged value of output, $X^a_4 = Q^a_{GN-1}$, the equation is estimated using two relative price variables. The first is the producer price of groundnuts relative to the aggregate price level in the nonagricultural sector, written as

$$X^a_2 = P^a_{GN}/P^a_{NA}, \tag{15}$$

with P^a_{GN} representing, as before, the nominal producer prices of groundnuts and P^a_{NA} the price in the nonagricultural sector.

The second relative price variable is that for producers in other agricultural sectors that are competing with groundnuts for production resources, P^a_{OA}. It is defined as

$$X^a_3 = P^a_{OA}/P^a_{NA}. \tag{16}$$

The relative price of food is used in the computations as a proxy for the incentives in competing agricultural activities. The general price level in the nonagricultural sector, P^a_{NA}, has nontradable and tradable components. The tradable component is affected by changes in trade regimes and other domestic policies, as is the price of groundnuts.

In the approach adopted here, the estimates of α_i and the values for X_i in the absence of policies affecting the groundnut sector are used to compute the levels of groundnut output that would obtain for the individual countries. The adjusted levels of output and the corresponding changes compared with actual levels can then be estimated:[11]

$$Q^e_{GN} = \Sigma_i \alpha_i X^e_i + \mu_{GN}, \tag{17}$$

$$q_{GN} = (Q^a_{GN} - Q^e_{GN})/Q^e_{GN}. \tag{18}$$

The superscripts a and e refer to actual and adjusted or equilibrium values, respectively, of the corresponding variables q_{GN} denotes the change in output levels.

Equation (17) gives the equilibrium level of groundnut output, Q^e_{GN} and equation (18) yields the changes in output due to changes in relative prices. The changes in exports by country can be computed by deducting from Q^e_{GN} the changes in domestic consumption of groundnuts resulting from the changes in relative prices. Due to data limitations and the resulting difficulties in obtaining satisfactory consumption parameter estimates, a simpler approach is used to approximate the changes in country groundnut exports: the actual annual shares of exports in groundnut output are assumed to remain unaffected by the changes in relative prices. This assumption is less restrictive in The Gambia and Senegal than it is in Sudan, since, in these two cases, exports have accounted for the largest part of the marketed share of the crop

---

[11]See Stryker (1990, 168) on the use of the entire original equation, including the residuals, instead of just the elasticity estimates in computing the effect of changes in relative prices.

during most of the period covered by the study. Accordingly, the following formula is used to approximate the changes in exports for each year:

$$q_x = k(Q_{GN}^a - Q_{GN}^e)/Q_x^a. \tag{19}$$

The change in country export quantities is given by q_x, while Q_x^a denotes the actual level of exports where domestic policy distortions exist. The coefficient k is the share of exports in country groundnut output.

The changes in export revenues can be calculated by multiplying the absolute changes in exported groundnut quantities by the prevailing world market price. That is,

$$g_v = kP_{GN}^w(Q_{GN}^a - Q_{GN}^e)/Q_v^a. \tag{20}$$

The formulation assumes that the world market price, P_{GN}^w is not affected by changes in country exports resulting from the adjustment in domestic policies. Q_v^a and g_v are the actual levels of and changes in country groundnut export revenues.

In order to compute equations (18), (19), and (20) to obtain the changes in country groundnut output and exports, X_2^e in equation (17)—that is, the adjusted relative groundnut producer prices—must be estimated first. This is done following the approach developed in Krueger, Schiff, and Valdés 1988. Following equation (15), the adjusted relative prices for groundnuts can be written as

$$X_2^e = P_{GN}^e / P_{NA}^e. \tag{21}$$

P_{GN}^e stands for the adjusted groundnut producer prices and P_{NA}^e is adjusted non-agricultural prices. In the situation where there are no policy imbalances, direct or indirect, absolute producer prices can be expressed as

$$P_{GN}^e = E^e P_{GN}^w - T_{GN}^e, \tag{22}$$

where E^e represents the adjusted exchange rate that would prevail in the absence of policy imbalances, as derived in the next section, and T_{GN}^e designates the cost of transferring groundnuts from the producing areas to the ports for export.[12]

Similar to groundnut prices, prices for the tradable component in the nonagricultural price index will also change in response to the elimination of policy distortions. Based on Krueger, Schiff, and Valdés 1988, the resulting changes in the nonagricultural price level can be computed using the following expression:

$$P_{NA}^e = h(E^e/E^a)P_{NAT}/(1 + t_{NAT}) + (1 - h)P_{NAH}, \tag{23}$$

where E^a represents the actual official exchange rate before the adjustment in domestic policies and h is the share of tradables in the nonagricultural sector. t_{NAT} denotes the tariff equivalent of trade restrictions in the nonagricultural sector, and P_{NAT} and

---

[12]The transfer costs will normally change as a result of the adjustment in domestic policies. In computing the model for individual AGC countries, the costs are assumed to change with the price level in the nonagricultural sector. The adjusted transfer costs are therefore calculated as follows: $T_{GN}^e = T_{GN}^a (P_{NA}^e / P_{NA}^a)$, where the superscripts a and e refer to actual and equilibrium values.

P_{NAH} are the price levels for the tradable and nontradable components of the non-agricultural price index, respectively.

The last step in estimating equilibrium groundnut prices by country in equation (21) and the output and export effects in equations (18), (19), and (20) is to calculate the equilibrium exchange rate E^e that would prevail when domestic policies are adjusted. The model used for this purpose is presented in Appendix 2. It yields the following formula for the estimation of the equilibrium exchange rate:[13]

$$E^e = E^a \frac{B_a + \eta_d \left[tm/(1+tm)\right] m_d + \varepsilon_s \left[tx/(1-tx)\right] X_s}{\varepsilon_s X_s + \eta_d M_d} + E^a. \qquad (24)$$

Equation (24) gives the nominal equilibrium exchange rate, which is used in equations (22) and (23) to adjust price levels for each country. The variables are defined in Appendix 2.

Since the change in the country exchange rate affects domestic prices, the rate calculated in equation (24) must be adjusted by this effect to yield the real change in the exchange rate. For that purpose, the actual real exchange rate is defined as

$$E_r^a = E^a/P_{NA}. \qquad (25)$$

The adjusted real exchange rate that incorporates both the change in the nominal rate and in the overall price level is then

$$E_r^e = E^e/P_{NA}^e. \qquad (26)$$

The effects of domestic policies on the production and export of groundnuts by individual AGC countries can now be estimated. First, the equilibrium exchange rates for each country are calculated using equation (24). Second, the results are inserted into equations (22) and (23) to obtain the adjusted real groundnut prices for each country. Third, these adjusted prices are used with equation (17) to obtain the adjusted groundnut output and with equations (18) and (19) to derive the changes in country outputs and exports. The results of the estimations for The Gambia, Senegal, and Sudan and the variables used therein are discussed in the following sections.

Evidence of Retarded Growth in the Groundnut Sector

Changes in external factors, as reflected in the evolution of each country's terms of trade for groundnuts, have not been as influential in the decline of trade in that sector as it is often argued. In fact, the results of the decomposition of real groundnut export revenues for individual member countries and for the AGC as a whole have indicated that the effects of falling export volumes were far larger than the effects of changes in the terms of trade. Furthermore, the figures in Table 10 reveal that AGC

---

[13]The model is based on Krueger, Schiff, and Valdés 1988. For other applications see Stryker 1990; Intal and Power 1990; Jansen 1988; Jenkins and Lai 1989; Moon and Kang 1989; and Garcia and Llamas 1989.

countries have continuously lost market shares, while competing exporters from Asia and Latin American have increased their shares of world groundnut exports.

It seems, therefore, that the causes of the decline in groundnut trade in AGC countries are more domestic than external. The factors undermining growth in the groundnut sector are manyfold: sectoral and overall economic policies of the countries are among the most important. One group of policies that may have contributed to the decline of the groundnut industry is marketing and pricing policies.

Direct Effects of Sectoral Policies on Real Groundnut Prices. The direct effect is reflected in the wedges between actual relative groundnut producer prices P_{GN}^a/P_{NA}^a and relative export prices, calculated at the actual official exchange rate and in the presence of protection in favor of other sectors of the economy and adjusted for transfer costs $P_{GN}^{a'}/P_{NA}^a$.[14] That is,

$$(P_{GN}^a/P_{NA}^a)/(P_{GN}^{a'}/P_{NA}^a) - 1 = P_{GN}^a/P_{GN}^{a'} - 1. \qquad (27)$$

The first quotient on the left-hand side of equation (27) corresponds to the actual relative producer price as given in equation (15). The numerator in the second quotient is calculated as in equation (22), but with actual instead of equilibrium values of the exchange rate and transfer costs.

In the case of Senegal, where parastatals and state-owned mills have been given monopoly over groundnut exports, the transfer costs used in the computations are based on cost figures reported by these sources. Given the tendency of parastatals to inflate marketing costs and extract subsidies from the government, the official figures had to be corrected in order to obtain adequate estimates for the costs of transferring groundnuts to the ports of export. There is evidence that marketing parastatals have consistently inflated transfer costs by reporting two types of costs, "declared losses" and "miscellaneous fixed costs," which, according to the arrangement between the marketing institutions and the government, are entirely subsidized (Jammeh 1987; SOFRECO 1988a). Prior to the early 1980s when groundnut marketing became the responsibility of the exporting parastatal SONACOS, which was also overlooking the milling companies, the declared losses were negligible. After SONACOS took over, they climbed to an average as high as 17 percent of the reported costs. The miscellaneous fixed costs, on the other hand, amounted to 26 percent of total costs (SOFRECO 1988a, 97).

Transfer costs for the entire period were not available for The Gambia. Given the similarity between the two countries, the real costs of transferring groundnuts are assumed to be similar to those in Senegal.[15] In Sudan, the estimate for 1980 by El Bashir and Idris (1983) has been extrapolated by assuming that the change in transfer costs follows that of the price level in the service sector.

The proportional differences between the transfer-cost-adjusted border prices and actual country producer prices are used as indicators for the direct effect of sector

---

[14]Transfer costs refer to the costs of assembly, storage, transport, and other elements of marketing costs.

[15]Studies carried out recently in both countries have found cost levels in the two countries comparable (Kristjansen et al. 1990; Jones 1986).

policies on incentives in the groundnut sector.[16] The proportional differences correspond to the levels of direct nominal protection in the respective countries (Figure 9).

Three phases can be distinguished in the evolution of the level of direct nominal protection (Table 13). In the 1960s, when AGC countries were making a strong showing on world groundnut markets, domestic prices were very close to border price levels. The slightly positive protection observed during this period may reflect the subsidization of the state-run marketing systems, particularly in The Gambia and Senegal, as will be shown later in discussing the breakdown of country export revenues. During the period of the dramatic decline in groundnut production and exports in the 1970s, direct taxation increased in both countries, while the rate of positive protection climbed steeply in Sudan. Finally, during the 1980s, the level of taxation increased further in The Gambia, and Senegal joined Sudan in raising domestic prices significantly above export levels and subsidizing exports.

Despite the attempt to correct official transfer costs for operational inefficiencies and reporting inadequacies, the exclusive monopoly of parastatals over procurement,

Figure 9—Direct nominal protection of groundnuts in selected AGC countries, 1966-88

Percent

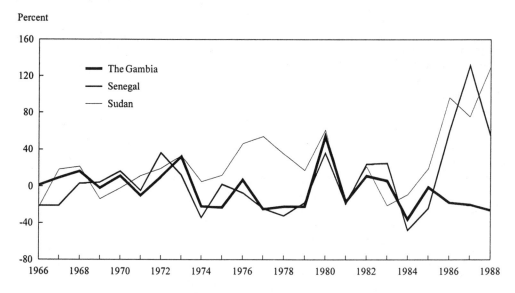

Source: Calculated from various sources used to compile Tables 1-3 of the data supplement to this report, available on request from the International Food Policy Research Institute.
Notes: Based on the proportional difference between the relative border-equivalent producer price and the actual relative export price. AGC is African Groundnut Council.

---

[16]Due to the lack of data, the impact of input subsidies or taxes is not included in computing the level of protection. This is not likely to affect the results greatly. The level of use of intermediate inputs is relatively low and the effect on policies probably negligible, compared with the effects on output prices. According to Jammeh (1987), for instance, input subsidies in Senegal averaged CFA 1.5 billion annually during the 16 years of the Agricultural Program that ended in 1980. The annual volume of production during this period was about 1 million tons on average.

Table 13— Direct effects of sector policies on real country groundnut prices, The Gambia, Senegal, and Sudan, 1966-88

	The Gambia			Senegal			Sudan		
Period	Actual Border Equivalent of Producer Price[a]	Actual Real Export Price[b]	Direct Price Effect[c]	Actual Border Equivalent of Producer Price[a]	Actual Real Export Price[b]	Direct Price Effect[c]	Actual Border Equivalent of Producer Price[a]	Actual Real Export Price[b]	Direct Price Effect[c]
	(D/metric ton)			(F/metric ton)			(£S/metric ton)		
1966-72	207.54	199.05	0.05	24,874.13	24,827.06	0.02	37.54	36.68	0.04
1973-80	237.34	257.00	-0.03	28,639.21	32,918.11	-0.08	47.38	36.47	0.33
1981-88	249.93	295.04	-0.13	24,330.04	23,654.31	0.25	52.59	41.97	0.37

Sources: Calculated from Tables 1-3 in the data supplement to this report, available on request from the International Food Policy Research Institute.
[a]The actual relative producer price is adjusted for transfer costs and export taxes.
[b]The nominal exchange rate × Groundnut export unit value / Actual nonagricultural price level.
[c](Actual border equivalent of relative producer price – Actual relative export price) / Actual relative export price.

processing, and exports, especially during the 1960s, raises the question of hidden taxation associated with these activities. For example, the breakdown of export prices in Table 14 shows relatively high levels of transfer cost shares for Senegal and The Gambia during the 1960s, followed by a continuous decline in the subsequent years. This, coupled with relatively stable producer shares of about 70 percent in the export price, indicates that some of the protection observed in that period went to subsidize parastatals rather than to benefit domestic groundnut producers.

It is also interesting to compare the differences in the proportions of the export price going to producers in the three countries and their changes over time. Initially at a much higher level than in the other two countries, the ratio of the producer price to the export price of groundnuts in Sudan increased continuously, reflecting a faster increase in Sudanese producer prices than in actual export prices and, consequently, a rising level of price support in that country. Sudan could sustain this high level of protection because more than 90 percent of production was consumed domestically (see Tables 4 and 8). At the same time, the small share of production that was exported was subsidized through various measures, one of which was the administration of preferential exchange rates for groundnut exports (El Bashir and Idris 1983; Louis Berger International 1983).

In The Gambia and Senegal, nominal producer prices kept up with world market prices throughout the 1970s, whereas the share of transfer costs fell by almost a third. These developments indicate that the hidden subsidization of marketing institutions during the 1960s had reverted to open taxation, as reflected in the sums of the two ratios, which in both cases are less than unity. Open taxation continued into the 1980s in The Gambia, whereas producer prices in Senegal were allowed to rise slightly above their export price levels.

With the exception of The Gambia, and except for the 1970s, sectoral policies do not seem to have suppressed producer prices in the groundnut sector very much. However, sectoral policies are not the only, and in many instances not the most important, type of policy measures that affect sectoral incentives. As stated earlier, imbalances in economy-wide policies can be more detrimental to the performance of tradable sectors such as groundnuts. In the end, it is the combination of both direct and indirect policy effects that determines the incentive environment of the groundnut sector.

Table 14— Ratios of producer prices and transfer costs to groundnut export prices, The Gambia, Senegal, and Sudan, 1966-88

Period	The Gambia		Senegal		Sudan	
	Transfer Costs/ Export Price[a]	Producer Price/ Export Price[b]	Transfer Costs/ Export Price[a]	Producer Price/ Export Price[b]	Transfer Costs/ Export Price[a]	Producer Price/ Export Price[b]
1966-72	0.32	0.73	0.34	0.67	0.26	0.77
1973-80	0.23	0.74	0.24	0.68	0.40	0.87
1981-88	0.17	0.71	0.23	1.03	0.24	1.12

Sources: Calculated from Tables 4-6 in the data supplement to this report, available on request from the International Food Policy Research Institute.

[a]Actual transfer costs / (Groundnut export unit value × Nominal exchange rate).

[b]Actual nominal producer price / (Groundnut export unit value × Nominal exchange rate).

Aggregate Effects of Domestic Policies on Real Groundnut Prices. In this section, the indirect effect of domestic policies outside of the groundnut sector is combined with the direct effect of the sectoral policies to show the aggregate impact of country policies on prices and incentives in the groundnut sector. The indirect effect captures the impact of macroeconomic and trade policies on country real exchange rates and that of protection to importing nonagricultural sectors. It can be seen from the derivation of the real exchange rate model that the indirect effect is made up of three different components. The first is the adjustment in country real exchange rates that would be necessary to eliminate the unsustainable imbalances in country current accounts. The second component is the adjustment in real exchange rates that would result from the removal of trade restrictions. The final component is the reduction in the protection of nonagricultural prices relative to country groundnut prices that would follow from the elimination of trade restrictions.

Accordingly, the first step in calculating the aggregate effect of domestic policies on the groundnut sector is to estimate the adjustment in real exchange rates that would follow from eliminating the imbalances in country current accounts and removing trade restrictions. This is done by computing equations (24) and (26). Besides the levels of import and export taxes and the current account deficit, which are readily available from national statistics, the data needed for the computations include the elasticities of demand for and supply of foreign exchange with respect to the real exchange rate.

Given that AGC countries are price takers, both demand for their exports and the supply of their imports are infinitely elastic. Therefore, the elasticities of supply and demand for foreign exchange converge toward the elasticities of export supply and import demand respectively (see Intal and Power 1990, 313). In computing equation (24) the estimate by Khan (1974) of 1.1 for the elasticity of import demand for Ghana is used for all three countries. For the elasticity of export supply, a value of 0.6 is used, which corresponds to about one-half of the estimate by Bond (1983) for the aggregate export of agricultural raw material by African countries. The inability to obtain empirical estimates for the elasticities should not compromise the results of the model, since the indirect effects are not expected to be very sensitive to changes in their values (Krueger, Schiff, and Valdés 1988, 260).

The level of disequilibrium in the real exchange rates caused by macroeconomic policy imbalances and trade restrictions in AGC countries can be seen in Figure 10. The figure shows that country exchange rates have been held consistently over their equilibrium levels during the entire period covered by the study. The changes that have occurred in each subperiod are shown in Table 15. For each country, columns 1 and 2 correspond to E^e and E_r^e in equations (24) and (26), respectively. The numbers indicate that, in real terms, actual exchange rates in The Gambia, Senegal, and Sudan have diverged from their equilibrium levels by 20-45 percent over the study period, and that the level of disequilibrium has continuously increased through all three decades. The real exchange rate overvaluation was the highest in Senegal—at 30 percent or more—during the first two periods, and it stayed at that level in the last period. In contrast, the real exchange rate appreciated much more rapidly in the other two countries. By the 1980s, the rate of appreciation in The Gambia had reached that of Senegal. However, the increase was much more dramatic in Sudan, where the rate of appreciation soared to 45 percent.

The total effect of domestic policies on incentives to production and trade in the groundnut sector can now be estimated by comparing actual relative groundnut prices

41

Figure 10—Degree of divergence from the equilibrium exchange rate

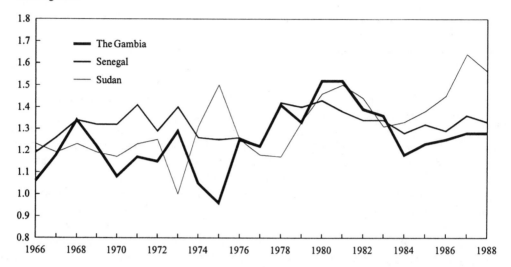

Source: Calculated from various sources used to compile Tables 7-12 of the data supplement to this report, available on request from the International Food Policy Research Institute.

Note: The equilibrium exchange rate is the level that would have kept the country's current account balance at a sustainable level.

Table 15—Exchange rate disequilibrium in The Gambia, Senegal, and Sudan, 1966-88

	The Gambia		Senegal		Sudan	
Period	Nominal[a]	Real[b]	Nominal[a]	Real[b]	Nominal[a]	Real[b]
1966-72	1.16	1.17	1.29	1.30	1.15	1.21
1973-80	1.28	1.25	1.37	1.33	1.25	1.28
1981-88	1.33	1.31	1.37	1.33	1.49	1.45

Sources: Calculated from Tables 7-9 in the data supplement to this report, available on request from the International Food Policy Research Institute.

[a] Nominal divergence = Nominal equilibrium exchange rate/Nominal actual exchange rate (E^e/E^a).

[b] Real divergence = Real equilibrium exchange rate/Real actual exchange rate

$$\left\{ \frac{E^e}{P^e_{NA}} \middle/ \frac{E^a}{P^a_{NA}} \right\}$$

where superscripts a and e refer to actual and equilibrium values. P^e_{NA} is the adjusted nonagricultural price level, as expressed in equation (23). E^e is the equilibrium exchange rate, defined in equation (24).

42

to the relative prices that would prevail at the equilibrium exchange rate. This is done by computing expression (28) below, which calculates the proportional difference between the adjusted border-equivalent producer price (equation 7) and the export price evaluated at the equilibrium exchange rate (equation 13):

$$(P_{GN}^a/P_{NA}^a)/(P_{GN}^e/P_{NA}^e) - 1. \tag{28}$$

The adjusted groundnut producer and nonagricultural prices, P_{GN}^e and P_{NA}^e are as defined in equations (22) and (23). The results of the computations are plotted in Figure 11. Changes in the level of total protection contrast starkly with the direct effects shown in Figure 9. Except for The Gambia, the direct effects appear to have been detrimental mainly during the 1970s. In Sudan, groundnuts even enjoyed strong direct protection. In contrast, total protection to domestic groundnut sectors resulting from country sectoral as well as overall macroeconomic policies and trade regimes was extremely negative for most of the period covered by the analysis. Hence, the indirect effects emanating from policies outside of the groundnut sector have in all three countries exacerbated or overcompensated for the impact of policies targeted directly to that sector.

The roles played by country trading regimes and macroeconomic policies through their impact on the real exchange rate in reducing incentives in the groundnut sector were overwhelming (Table 16). The implicit taxation induced at that level exceeded the positive protection granted to the sector through direct policy measures during the 1960s. This resulted in net taxation levels of 10-22 percent in the three

Figure 11—Total nominal protection of groundnuts in The Gambia, Senegal, and Sudan, 1966-88

Percent

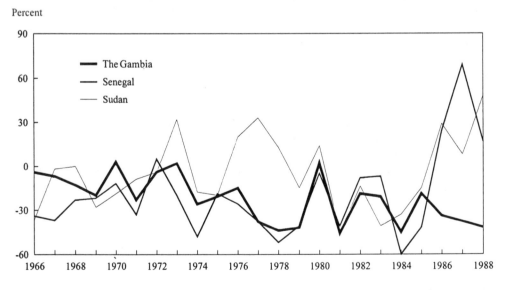

Source: Calculated from various sources used to compile Tables 13-18 of the data supplement to this report, available on request from the International Food Policy Research Institute.
Note: Indicates the proportional difference between the actual relative producer price and the relative export price at the equilibrium rate and without trade restrictions.

Table 16— Direct and aggregate effects of country policies on groundnut prices in The Gambia, Senegal, and Sudan, 1966-88

Period	The Gambia		Senegal		Sudan	
	Direct Price Effect[a]	Total Price Effect[b]	Direct Price Effect[a]	Total Price Effect[b]	Direct Price Effect[a]	Total Price Effect[b]
	(percent)					
1966-72	0.05	-0.10	0.02	-0.22	0.04	-0.14
1973-80	-0.03	-0.23	-0.08	-0.31	0.33	0.07
1981-88	-0.13	-0.33	0.25	-0.06	0.37	-0.08

Sources: Calculated from Tables 13-15 in the data supplement to this report, available on request from the International Food Policy Research Institute.

[a](Actual border equivalent of relative producer price − Actual relative export price)/Actual relative export price.

[b](Actual border equivalent of relative producer price − Relative equilibrium export price)/Relative equilibrium export price.

countries. In the following decade, the situation in Sudan was reversed through a rapid increase in the level of direct protection. But, in The Gambia and Senegal, the level of direct and indirect disincentives facing groundnut production and trade increased tremendously during the 1970s, a period when the world groundnut economy was booming (see Figures 5 and 6). It is interesting to note that this "boom" period was the time when the AGC countries suffered steep losses in market shares (see Figure 8). Finally in the 1980s, the level of taxation in Senegal fell sharply as the result of a substantial increase in the level of direct protection, while the situation in the other two countries worsened due to the deterioration of the direct policy environments in The Gambia and the indirect one in Sudan.

Effects of Country Policies on Groundnut Production and Exports. In order to calculate the impact of domestic policies on country production and export of groundnuts, equation (14) first has to be estimated to yield the elasticity parameters, the α_i's, for the three countries. In carrying out the estimation, it is assumed that food crops, such as millet and sorghum, are the main potential competitors for groundnuts. In Sudan, groundnuts may also compete with wheat in the irrigated area of Gezira. However, only about one-third of groundnuts are produced in that area, and the allocation of land to the different crops is determined more by tenancy size and government measures than by prices.[17] Thus, the estimation of the output equation for Sudan is based on data from the rainfed areas of Kordofan and Darfur, which account for two-thirds of groundnut production, and where groundnuts compete mainly with millet and sorghum.

In all three cases, the producer price of millet/sorghum relative to nonagricultural prices is therefore chosen to represent the price level in competing agricultural subsectors, as expressed in equation (16). The estimated functions for the three countries, which are based on the data contained in the supplement to this report, are as follows:[18]

---

[17]Various attempts to estimate supply equations including data from this area did not yield significant parameters for the price variables.

[18]The specification of the output functions does not take into consideration the role of inputs, due to lack of data on inputs (see footnote 16).

For The Gambia,

$$\ln Q_{GN} = -4.18 + 0.41 \ln P_{GN}/P_{NA} + 0.47 \ln P_{OA}/P_{NA} + 0.99 \ln Q_{GN-1}; \qquad (29)$$
$$\phantom{\ln Q_{GN} = } (2.03) \ (2.51) \qquad\qquad (2.64) \qquad\qquad (5.86)$$

$\overline{R}^2 = 0.60; H = 0.67.$

For Senegal,

$$\ln Q_{GN} = -3.65 + 0.62 \ln P_{GN}/P_{NA} + 0.66 \ln Q_{GN-1}; \qquad (30)$$
$$\phantom{\ln Q_{GN} = } (1.11) \ (2.25) \qquad\qquad (3.81)$$

$\overline{R}^2 = 0.40; H = 1.07.$

For Sudan,

$$\ln Q_{GN} = 2.70 + 0.42 \ln P_{GN}/P_{NA} - 0.20 \ln P_{OA}/P_{NA} + 0.40 \ln Q_{GN-1}; \qquad (31)$$
$$\phantom{\ln Q_{GN} = } (2.0) \ \ (1.43) \qquad\qquad (1.05) \qquad\qquad (2.53)$$

$\overline{R}^2 = 0.27; H = 1.59.$

The numbers in parentheses are t-values.

The estimated own-price elasticity coefficients have the correct sign and are significant at the 0.05 level for The Gambia and Senegal, and at a weaker 0.20 level for Sudan. The coefficients for the lagged output variable are strongly significant and have the correct sign as well. The estimates for the coefficient of food sector price are significant only in The Gambia. They are positively signed, indicating a complementary relationship in production between groundnuts and the two food crops, millet and sorghum.

Following equation (17), the level of output that would prevail in the absence of the disincentives created by domestic policies and evidenced in the foregoing section can be calculated by (1) inserting the equilibrium values of the independent variables, the X_i^e into the original equations (29 to 31) to obtain the predicted values, and (2) adding the residuals η_{GN} assuming the stochastic variation is not altered by changes in policies.[19] The independent variable of main interest is the adjusted relative producer price for groundnuts in individual countries. Apart from adjusting to the change in policy-induced taxation, relative producer prices will also change to reflect changes in transfer costs, to the extent that the latter are affected by policy changes. Accordingly, equilibrium producer prices are calculated as explained in equations (21) and (22).

Following the approach taken by Stryker (1990), the equilibrium output levels are computed for the short and long run. In the short run, the equilibrium relative prices and the lagged values of actual output are inserted into equations (29) to (31) and the residuals added to yield the adjusted groundnut output levels. The equilibrium output

---

[19]On the merits of using the entire original equations rather than the estimated elasticities, see Stryker 1990, 168.

45

levels in the long run are computed to capture the secondary effects of policy adjustment on current output resulting from changes in previous output levels. For this purpose, the actual lagged values of output in the short-run equations are replaced by the expected values computed from the estimated supply equations. The calculated divergences between actual and equilibrium groundnut output levels in the short and long run are plotted in Figures 12 and 13.

Furthermore, assuming that the effect of the changes in real prices on the shares of exports in groundnut output is negligible, the changes in groundnut export revenues can be estimated as shown in equations (19) and (20). The obtained changes in export revenues for the short and long terms are shown in Figures 14 and 15.

According to the results obtained, domestic policies have had a substantially negative impact on country groundnut production and exports. Except for Sudan for a few years and for the last two to three years of the period, domestic policies kept country production and exports well below the levels that would have prevailed in the absence of the disincentives caused by sector and economy-wide policies and trade regimes. The effects have been particularly strong in Senegal and The Gambia.

The magnitudes of the effects for the different time periods are shown in Tables 17 and 18. During the 1960s, actual country production was as low as 6-35 percent on average below equilibrium levels, due to the combined effect of domestic policies within and outside the groundnut sector. Senegal's output was depressed the most; its actual output levels were as much as 20-35 percent below the equilibrium levels over all three time periods.

Figure 12—Short-term divergence between actual and equilibrium levels of groundnut output, 1967-88

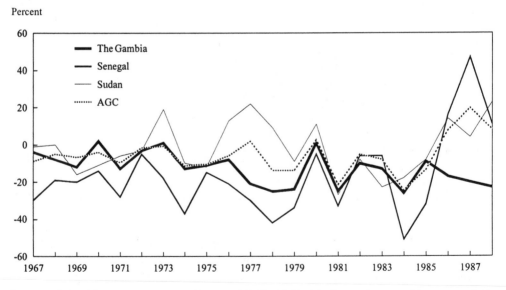

Source: Calculated from various sources used to compile Tables 22-24 of the data supplement to this report, available on request from the International Food Policy Research Institute.
Note: Indicates the proportional difference between the actual levels of output and the level of output adjusted for the effect of domestic policies. The values for African Groundnut Council (AGC) countries are weighted sums of country changes based on individual shares in total AGC output.

Figure 13—Long-term divergence between actual and equilibrium levels of groundnut output, 1967-88

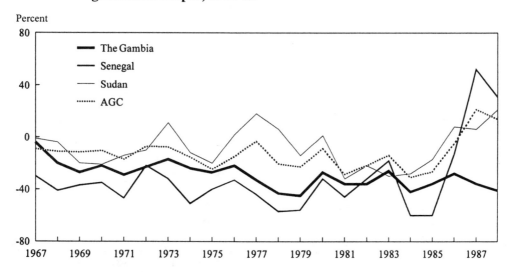

Source: Calculated from various sources used to compile Tables 22-24 of the data supplement to this report, available on request from the International Food Policy Research Institute.

Note: Indicates the proportional difference between the actual levels of output and the level of output adjusted for the effect of domestic policies. The values for African Groundnut Council (AGC) countries are weighted sums of country changes based on individual shares in total AGC output.

Figure 14—Short-term divergence between actual and equilibrium levels of groundnut export revenue, 1967-88

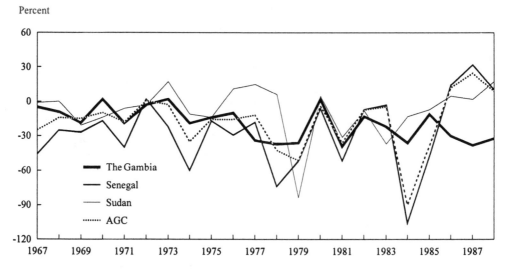

Source: Calculated from various sources used to compile Tables 22-24 of the data supplement to this report, available on request from the International Food Policy Research Institute.

Notes: Indicates the proportional difference between the actual level of export revenue and the level of revenue adjusted for the impact of domestic policies. The values for African Groundnut Council (AGC) countries are weighted sums of country changes based on individual country shares in total AGC exports.

47

Figure 15—Long-term divergence between actual and equilibrium levels of groundnut export revenues, 1967-88

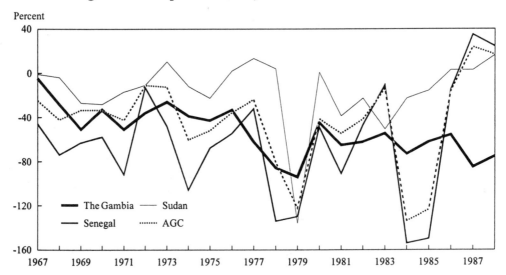

Source: Calculated from various sources used to compile Tables 25-27 of the data supplement to this report, available on request from the International Food Policy Research Institute.
Note: Indicates the proportional difference between the actual levels of export revenue and the level of revenue adjusted for the impact of domestic policies. The values for African Groundnut Council (AGC) countries are weighted sums of country changes based on individual shares in total AGC output.

Reflecting the worsening policy environment of the 1970s, the discrepancy between actual and equilibrium groundnut output expanded significantly throughout that decade both in Senegal and The Gambia. In contrast, the strong increase in direct policy support discussed earlier in Sudan helped keep its output close to its long-term equilibrium level, despite major imbalances in macroeconomic policies and trade regimes. However, by the end of the 1980s, the situation had worsened in The

Table 17— Estimated average annual change in groundnut output as a result of policies in AGC countries, 1967-88

Period	Short-Term Output Effect				Long-Term Output Effect			
	The Gambia	Senegal	Sudan	AGC[a]	The Gambia	Senegal	Sudan	AGC[a]
	(percent)							
1967-72	−6.33	−19.33	−6.17	−6.05	−20.83	−35.33	−11.67	−11.02
1973-80	−12.50	−25.25	5.38	−6.49	−29.75	−43.13	−1.00	−14.71
1981-88	−17.88	−6.75	−5.38	−4.17	−35.13	−18.25	−11.75	−11.46

Source: Calculated from Tables 22-24 in the data supplement to this report, available on request from the International Food Policy Research Institute.
Notes: Indicates the proportional difference between the actual levels of output and the level of output adjusted for the effect of domestic policies. The values for African Groundnut Council (AGC) countries are weighted sums of country changes based on individual shares in total AGC output.
[a]Weighted averages based on country shares in AGC output.

Table 18— Effects of domestic policies on groundnut exports, 1967-88

Country/Period	Actual Groundnut Export Revenue (1)	Groundnut Export Unit Value (2)	Short-Term Divergence from Equilibrium Export (3)	Long-Term Divergence from Equilibrium Export[a] (4)	Short-Term Export Revenue Effect[b] (5)	Long-Term Export Revenue Effect[b] (6)
	(US$ million)	(US$/metric ton)	(1,000 metric tons)		(percent)	
The Gambia						
1967-72	5.23	112.17	-4.07	-15.24	-8.8	-34.0
1973-80	10.85	276.75	-7.13	-20.23	-18.3	-53.5
1981-88	6.79	309.13	-6.43	-15.04	-27.6	-66.6
Senegal						
1967-72	50.20	103.68	-107.37	-250.12	-25.5	-57.7
1973-80	117.99	263.70	-141.49	-320.41	-34.8	-77.8
1981-88	65.91	229.58	-35.89	-108.33	-20.0	-51.1
Sudan						
1967-72	16.18	150.07	-6.68	-14.14	-7.5	-14.7
1973-80	58.14	289.40	1.32	-14.63	-6.9	-17.6
1981-88	21.39	352.23	-5.81	-11.34	-9.0	-16.3

Sources and notes: Groundnut oil exports (column 1) and export unit values (column 2) are from Food and Agriculture Organization of the United Nations, *FAO Trade Yearbook* (Rome: FAO, various years); columns (3) and (4) are deducted from estimated short- and long- term output effects by assuming the same shares of export in groundnut output as in the pre-equilibrium situation; columns (5) and (6) are calculated using equation (20) and, respectively, columns (3) and (4) for the value of $(Q^a_{GN} - Q^e_{GN})$, where Q^a_{GN} is the actual level of groundnut output in each country and Q^e_{GN}, the equilibrium level.

[a] In-shell groundnut equivalent.
[b] Average annual changes.

Table 19— Estimated average annual change in aggregate groundnut exports as a result of AGC country policies, 1967-88

Period	Short-Term Quantity Effect	Long-Term Quantity Effect	Short-Term Revenue Effect	Long-Term Revenue Effect
	(percent)			
1967-72	−5.50	−13.17	−13.47	−31.32
1973-80	−12.63	−31.00	−22.28	−53.66
1981-88	−9.50	−27.50	−16.39	−43.05

Sources: Based on results in Table 18 and country shares of African Groundnut Council (AGC) exports.
Note: The figures indicate the proportional difference between actual levels of export quantity and revenues and the level of exports adjusted for the effects of domestic policies. The AGC values are weighted sums of country changes (Table 18), based on individual country shares in AGC exports.

Gambia and Sudan. At the aggregate AGC level, the decline of output in all three countries contributed to a fall in total groundnut production of 5-15 percent.

The reduction in country output translated into considerable losses in groundnut exports. The decline in the volume of exports attributable to domestic policies was particularly strong in Senegal, the main exporting AGC member (Table 18). The actual quantity of groundnuts exported by Senegal at times fell more than 300,000 metric tons (unshelled equivalent) below their equilibrium levels, due to the disincentives created by domestic policies. Smaller absolute changes took place in Sudan and The Gambia, which exported much less than Senegal.

At the prevailing export prices, the reduction in export quantities meant export revenue losses of 20-70 percent for Senegal and The Gambia. For Sudan, the computed revenue losses vary from 10 to 20 percent. Translated into changes in aggregate AGC exports, these losses correspond to a decline of up to 30 percent in quantity and 54 percent in export revenues (Table 19).

These results clearly indicate that domestic policies contributed significantly to the decline of the groundnut sector of AGC countries and to a significantly larger extent than factors related to international groundnut and oilseed markets. They had strong, detrimental effects, both directly and indirectly, on prices and incentives in that sector. They suppressed producer prices directly and caused country real exchange rates to appreciate significantly. The ultimate consequence has been a substantial reduction in output, export volumes, and export revenues in individual member countries and the AGC as a whole.

6

THE GROUNDNUT DEMAND OUTLOOK
AND THE POTENTIAL ROLE OF
REGIONAL MARKETS

There was some indication in the discussion of trends in world trade in oilseeds in Chapter 4 that future global groundnut demand is more likely to see a shift in the location of import markets than a change in quantities. This chapter, therefore, starts with a review of the main import markets for groundnut products and turns to the consumption patterns of vegetable oils in selected regions and countries and groundnut import projections for the rest of the decade. In the second part of the chapter, the determinants of oilseed import demand in regional markets are analyzed and implications for future AGC exports to these markets are drawn.

Demand Outlook for Oilseed Products

European markets have traditionally been the most important import markets for oilseed products (Table 20). Their average share of demand in almost every oilseed product category between 1982 and 1987 was considerably higher than the combined share of all other importers. During the period, European markets accounted for 50-60 percent of world imports of unprocessed groundnuts, 80 percent of groundnut oil imports, and nearly 90 percent of groundnut cake imports. Despite the still-high levels of demand for groundnut products in European markets, demand is tilting rapidly toward other oilseeds. With the exception of sunflower and soybean oils, import growth rates for other oilseed products rank between 5 and 11 percent, compared with -5 to 0 percent for groundnut products.

The erosion of groundnut demand in Europe contrasts sharply with the strong import growth rates in the historically less important markets of Africa and Asia for groundnut oil and of North and Central America and South America for unprocessed groundnuts.

The percentage share of vegetable oils and fats in average daily per capita calorie intake in various countries at different levels of income is presented in Table 21. Since most vegetable oils are considered luxury goods, the demand for them is highly income elastic. However, where incomes are high and per capita calorie intake levels are in excess of 3,000 calories per day, the demand for vegetable fats and oils tends to taper off, but the absolute demand may well depend on taste and cultural preferences, as evidenced by the relatively high levels of consumption in Italy and the United States, compared with the moderate intake in France and Switzerland.

The direction that growth of demand is likely to take is shown in Table 22. Growth in per capita consumption of major fats and oils over the last 30 years has been more rapid in the subset of developing countries in the sample than in the group

Table 20— World imports of major oilseeds and oleaginous fruit products, by region, 1961-87

Region/Crop	Average Imports, 1982-87		Average Annual Growth Rate, 1961-87
	Quantity	Share	
	(1,000 metric tons)	(percent)	(percent)
World			
Soybeans			
Beans	27,294.64	100.00	7.99
Oil	3,665.91	100.00	9.42
Cake	22,770.44	100.00	11.98
Groundnuts			
In shell	105.16	100.00	1.15
Shelled	746.77	100.00	-2.99
Oil	387.35	100.00	0.10
Cake	620.31	100.00	-4.20
Palm			
Kernels	115.85	100.00	-7.89
Kernel oil	587.37	100.00	9.26
Kernel cake	833.51	100.00	6.66
Palm oil	4,712.47	100.00	10.52
Sunflowers			
Seeds	2,065.95	100.00	10.48
Oil	1,636.18	100.00	6.92
Cake	1,494.11	100.00	5.47
Africa			
Soybeans			
Beans	55.30	0.20	13.59
Oil	388.87	10.61	8.64
Cake	692.50	3.04	37.22
Groundnuts			
In shell	0.59	. . .	-10.35
Shelled	26.77	3.58	1.98
Oil	21.80	5.63	1.35
Cake	4.40	0.71	-3.65
Palm			
Kernels	0.02	. . .	-27.87
Kernel oil	42.58	7.25	12.56
Kernel cake	n.a.	n.a.	n.a.
Palm oil	703.23	14.92	18.84
Sunflowers			
Seeds	22.45	1.09	21.35
Oil	355.42	21.72	12.17
Cake	3.00	0.20	12.85
Asia			
Soybeans			
Beans	8,405.99	30.80	7.03
Oil	1,663.47	45.38	11.96
Cake	2,078.24	9.13	23.73
Groundnuts			
In shell	20.87	19.84	0.85
Shelled	188.13	25.19	4.74
Oil	48.56	12.54	1.96
Cake	67.00	10.80	1.29
Palm			
Kernels	16.40	14.16	-5.22
Kernel oil	64.60	11.00	23.12
Kernel cake	16.90	2.03	-51.02[a]
Palm oil	2,942.66	62.44	16.48

(continued)

Table 20 —Continued

Region/Crop	Average Imports, 1982-87		Average Annual Growth Rate, 1961-87
	Quantity	Share	
	(1,000 metric tons)	(percent)	(percent)
Sunflowers			
Seeds	18.01	0.90	-1.55
Oil	144.69	8.84	6.85
Cake	33.50	2.24	116.80[b]
Europe			
Soybeans			
Beans	15,204.46	55.70	11.28
Oil	762.86	20.81	-8.15
Cake	16,999.36	74.65	3.88
Groundnuts			
In shell	67.23	63.94	0.89
Shelled	391.35	52.41	-5.31
Oil	308.37	79.61	0.06
Cake	544.58	87.79	-4.46
Palm			
Kernels	98.37	84.91	10.62
Kernel oil	328.52	55.93	9.35
Kernel cake	816.47	97.75	3.84
Palm oil	913.86	19.39	6.47
Sunflowers			
Seeds	1,448.02	70.09	5.11
Oil	617.25	37.22	0.69
Cake	1,385.72	92.74	-0.68
South America			
Soybeans			
Beans	639.28	2.30	17.82
Oil	420.39	11.47	11.96
Cake	648.18	2.85	39.19
Groundnuts			
In shell	9.30	8.84	10.64
Shelled	2.37	0.32	7.75
Oil	0.27	0.07	-12.57
Cake	...	...	...
Palm			
Kernels	0.004	...	n.a.
Kernel oil	3.46	0.60	2.56
Kernel cake	0.12	...	n.a.
Palm oil	0.63	...	-16.51
Sunflowers			
Seeds	0.84	...	14.88
Oil	89.28	5.46	18.37
Cake	0.12	...	n.a.
North and Central America			
Soybeans			
Beans	1,659.56	6.08	6.86
Oil	208.16	5.68	9.97
Cake	1,004.26	4.41	7.16
Groundnuts			
In shell	7.05	6.70	9.55
Shelled	71.64	9.59	2.02
Oil	6.40	1.65	-1.74
Cake	...	...	...

(continued)

53

Table 20 —Continued

Region/Crop	Average Imports, 1982-87		Average Annual Growth Rate, 1961-87
	Quantity	Share	
	(1,000 metric tons)	(percent)	(percent)
Palm			
Kernels	0.98	0.80	27.65
Kernel oil	140.74	23.96	5.52
Kernel cake	0.004	. . .	n.a.
Palm oil	205.48	4.36	9.98
Sunflowers			
Seeds	572.92	28.84	27.25
Oil	177.03	10.82	6.22
Cake	63.00	4.22	18.28[c]

Sources: Food and Agriculture Organization of the United Nations, *FAO Trade Yearbook* (Rome: FAO, various years).
Notes: Shares may sum to less than 100 because Eastern European and Oceanic regions are not included. The ellipsis (. . .) indicate a nil or negligible amount. n.a. means data were not available.
[a]1976-87 only.
[b]1979-87 only.
[c]1977-87 only.

Table 21— Share of vegetable oils and fats in daily calorie intake, selected countries, 1979-81

Country	Total Calories per Capita per Day	Calories in Vegetable Oils and Fats per Day	Share of Total Caloric Consumption
			(percent)
Argentina	3,164	320	10.1
Bangladesh	1,837	40	2.2
Brazil	2,533	152	6.0
China	2,402	65	2.7
Côte d'Ivoire	2,567	231	9.0
France	3,260	350	10.7
Ghana	1,746	117	6.7
India	2,056	130	6.3
Indonesia	2,367	143	6.0
Italy	3,434	499	14.5
Malaysia	2,422	220	9.1
Nigeria	2,327	269	11.6
Pakistan	2,180	168	7.7
Philippines	2,377	85	3.6
Senegal	2,339	293	12.5
Switzerland	3,259	331	10.2
United States	3,455	491	14.2
U.S.S.R.	3,207	202	6.3
Zaire	2,097	159	7.6

Source: World Bank, *Food Products, Fertilizers, and Agricultural Raw Materials*, vol. 2 of *Price Prospects for Major Primary Commodities* (Washington, D.C.: World Bank, 1988).

Table 22— Trends in per capita consumption of major fats and oils, selected countries, 1960-87

Country	1960	1970	1980	1987	Average Annual Change
		(kilograms/capita/year)			(percent)
Brazil	5.45	8.04	16.95	18.70	9.0
China	2.46	2.65	4.65	7.50	7.6
Côte d'Ivoire	3.88[a]	6.22	9.64	9.34[b]	9.4
European Community	26.75	31.57	35.22	38.50	1.6
Ghana	4.86[a]	5.66	5.30	6.62[b]	2.4
India	5.75	5.48	6.54	7.20	0.9
Indonesia	2.58	3.08	6.02	8.70	8.9
Japan	6.95	13.13	15.81	19.30	6.6
Nigeria	9.19[a]	9.54	10.84	9.42[b]	0.2
United States	28.55	32.60	34.79	39.40	1.4
U.S.S.R.	13.64	17.71	20.52	22.80	2.4
World	9.53	10.70	12.73	13.10	1.4

Source: World Bank, *Food Products, Fertilizers, and Agricultural Raw Materials*, vol. 2 of *Price Prospects for Major Primary Commodities* (Washington, D.C.: World Bank, 1988).

[a]1961
[b]1986

of industrialized countries. India, where consumption has grown very slowly, and Japan, where it has grown rapidly, are the exceptions. During the period 1972-85, per capita consumption of oils and fats in developing countries grew by 28 percent, rising from an annual average of 5.0 kilograms in 1972-74 to 6.4 kilograms in 1985. Developed countries as a whole had a 5 percent annual growth rate and consumption levels rising from 19.9 kilograms in 1972-74 to 21 kilograms in 1985. Both income and population growth were faster in developing countries, which generally accelerated the demand for food. This was particularly true for vegetable oils and fats because they constitute relatively cheap and concentrated sources of both energy and protein (Watt and Merrill 1963).

Moreover, growing health concerns in industrial countries over the consumption of animal fats and oils are likely to encourage demand for mono-unsaturated vegetable fats such as groundnut oil. And, as the democratization process in Eastern European countries proceeds and their economies move to reflect market forces, demand there is also likely to shift from animal fats and oils to the relatively cheaper vegetable fats and oils. These developments, coupled with rapid population growth and rising incomes in developing countries, indicate that global demand for oilseeds will continue to grow for the next decade.

Turning to projections of the regional distribution of growth in import demand for groundnut products (Table 23), import demand in the European Community and other Western European countries is projected to grow by approximately 2 percent annually, despite the decline in the 1970s and early 1980s. The strongest growth in demand is expected to take place in former nonmarket economies and in developing African and South American countries. Total world imports and those of developed countries, however, are expected to grow only half a percentage point a year until the end of the decade. The stagnating trends in world groundnut demand are, therefore, not likely to

Table 23— Projected groundnut imports by economic regions

Economic Region	Imports			Growth Rate[a]		
	1990	1995	2000	1961-86	1970-86	1987-2000
	(1,000 metric tons)			(percent)		
Developed countries	439	510	544	0.1	−1.4	0.6
European Community (10 countries)	403	470	499	0.3	−1.3	1.6
Other Western European countries	36	40	45	0.4	−0.7	2.4
Nonmarket countries	5	6	8	−0.4	−0.8	5.9
Developing countries	152	173	197	0.6	−1.5	3.4
Asia	113	124	130	2.2	4.3	1.9
Africa	31	40	55	0.8	−1.0	8.1
Latin America and the Caribbean	8	9	12	−5.6	−20.1	5.5
World	596	689	749	0.2	−1.5	0.5

Sources: World Bank, *Food Products, Fertilizers, and Agricultural Raw Materials*, vol. 2 of *Price Prospects for Major Primary Commodities* (Washington, D.C.: World Bank, 1988).
[a]Least squares trend for the historical period, 1961-86; end point for the projected period 1987-2000.

change. This is an indication that future growth in international demand will not have any significantly stimulating effect on groundnut oil exports and that competition will increase considerably on international markets. However, given the strong expansion of demand in African countries depicted in this report and the projections presented here, regional markets may provide future outlets for AGC exports.

Regional Import Demand and AGC Groundnut Exports

Analyzing the role that regional markets have played in the past as a destination for AGC groundnut exports and the factors that drive regional import demand for oilseed products is necessary to understand the challenge AGC countries will face in expanding exports to these markets. Thus, a constant market share (CMS) model is applied to AGC groundnut exports to regional markets, followed by an econometric analysis of regional import demand for different vegetable oils and for groundnut oils from AGC countries.

The CMS model used to analyze the role of regional markets in AGC countries' trade in oilseed products isolates the contribution of three different factors to the change in export performance by individual AGC countries: (1) the competitiveness of individual AGC countries in different oilseeds markets; (2) the relative expansion of demand for individual oilseeds; and (3) the geographic orientation of country exports. (The model is described in Appendix 3). It decomposes the changes in country export shares into the three components. The first is the competitive effect (*CE*):

$$CE = \frac{(1 + g_i^m)}{(1 + g_i^w)},$$
(32)

where g stands for the growth rate of exports of oilseed i's exports by AGC country m and by the world w. This effect expresses the contribution of changes in the competitiveness of country m in a given oilseed market to the change in its overall trade performance. The effect is positive or negative depending on whether the ratio is greater or less than 1.0 for the country being considered.

The second component is the product effect (*PE*):

$$PE = \frac{(1 + g_i^w)}{(1 + g^w)} \frac{X_{it0}^m}{X_{t0}^m},$$ (33)

where X and X_i represent, respectively, the aggregate and individual oilseed exports by country m in period t_0. The product effect reflects the contribution of changes in single-product markets to the change in the aggregated market share of country m. The higher the share of products in the exports of a country that experiences faster growth than the world average, the greater will the product effect be.

The third component is the market effect (*ME*):

$$ME = \frac{(1 + g_j^w)}{(1 + g^w)} \frac{X_{jt0}^m}{X_{t0}^m},$$ (34)

where the index j denotes the regional markets, and the remaining variables and indices are as defined. This last component reflects the impact of the geographic orientation of country m's exports on the growth of its trade share. The effect is positive if the country directs a large share of its exports to markets that grow more rapidly than the world average.

In addition to singling out the past and potential role of regional markets, the model can also be used to test the external demand constraint hypothesis. For example, the external demand constraint would imply that g_i^w (the rate of growth of world demand) does not exceed g_i^m (the country export growth rate) or, equivalently, a competitive effect not below unity. Similarly, the product effect is helpful in highlighting the consequences of the change in composition taking place in world demand for vegetable oils, as discussed in Chapter 4 and earlier in this chapter. A low product effect, for instance, would reflect the effect of a slower growth of world demand for groundnut products compared with all oilseed products.

The model was calculated for individual AGC countries, focusing only on groundnuts for the competitive and product effects and on the African region for the market effect (Table 24). The competitive effect is negative (less than 1.0) for all AGC countries, which implies that AGC members have not been able to maintain their initial shares in world groundnut markets over the last three decades. The loss in competitiveness was substantial for both Niger and Nigeria, the latter with an initial trade share that exceeded 25 percent of total world groundnut exports (see Table 10). The negative effect means that AGC exports have been falling faster than global exports. This clearly contradicts the argument of an external demand constraint.

The product effect columns in Table 24 refer to the specific contribution of groundnuts to the change in the position of AGC countries in international oilseed and oilfruit product markets. With the exception of Nigeria and Sudan, groundnut products make up to 90 percent or more of AGC countries' exports of oilseed products. The growth rate of world groundnut exports, however, has been less than 50 percent of that for the aggregate of oilseed commodities. The stronger growth of nongroundnut oilseed products explains why a groundnut exporter could lose shares in the total

Table 24—Market share results for oilseed and oilfruit products, AGC countries, 1962-87

Country	Average Share 1962-67	Competitive Effect[a]	Product Effect		Market Effect	
			Relative Growth Rate[b]	Share[c]	Relative Growth Rate[d]	Share[e]
	(percent)					
The Gambia	0.4	0.845	0.406	0.839	2.523	n.a.
Mali	0.2	0.387	0.406	0.894	2.523	0.230
Niger	0.5	0.000	0.406	0.962	2.523	0.230
Nigeria	7.4	0.001	0.406	0.541	2.523	n.a.
Senegal	3.3	0.408	0.406	0.890	2.523	0.031
Sudan	1.9	0.510	0.406	0.367	2.523	n.a.

Sources: Food and Agriculture Organization of the United Nations, *FAO Trade Yearbook* (Rome: FAO, various years).
[a]The competitive effect is the growth rate of a country's exports of groundnut products relative to exports of other oilseed and oleaginous fruit products.
[b]Refers to the growth rate of a country's groundnut product exports relative to exports of other oilseed and oleaginous fruit products.
[c]Refers to the share of groundnut products in total exports of oilseed and oleaginous fruit products.
[d]Refers to the growth rate of African imports of oilseed and oleaginous fruit products compared with the growth rate of world imports.
[e]Refers to the share of Africa in each country's exports of oilseed and oleaginous fruit products.

vegetable oils markets, but not necessarily in groundnut markets. It is also important to note that the relative fall in demand for world groundnut products largely reflects the fall in exports from AGC countries, which still account for nearly 20 percent of world exports (Table 10). For example, the decline of AGC exports between the 1960s and the 1980s was three times larger than the reduction in world demand for groundnut oil of about 45,000 metric tons (Table 25).

The market effect (the last two columns in Table 24) indicates the contribution of African markets, as a destination for AGC countries' exports, to the change in their overall trade share. It shows that AGC countries seldom export to African markets: the African share in the exports of the largest exporter in the AGC, Senegal, is only 3 percent. Demand in African markets for all oilseed products, however, grew more than two-and-a-half times faster during the study period than demand in world markets. Furthermore, as shown in the previous chapter, the African market is one of the regions with a strong expansion of demand for groundnut products. It would seem, therefore, that a stronger orientation of AGC exports to regional (African) markets would stimulate export growth more than stagnating overseas markets. The question to ask now is if regional oilseed markets continue to expand as rapidly as in the past, to what extent and under which conditions could AGC exporters benefit from such an expansion?

The outlook section in the last chapter predicted that demand for groundnut products in African markets would increase by nearly 10 percent a year in the 1990s. These projections do not, however, take into consideration the demand for other oilseeds, which are also expected to expand considerably. Therefore, whether the fast-growing regional markets can help boost future AGC groundnut exports will depend primarily on the changes in the level and composition of regional vegetable oil demand and on the competitiveness of AGC exporters on these markets.

Table 25— Groundnut and palm oil imports and groundnut oil exports by West African and other African countries, 1961-87

Region/Commodity	1961-67	1973-77	1983-87
	(1,000 metric tons)		
World	**Vegetable Oil Imports**		
Groundnut oil	438.20	513.08	393.42
Palm oil	626.56	1,930.34	4,888.49
Other African countries			
Groundnut oil	10.95	23.08	22.42
Palm oil	23.18	110.02	718.60
West African countries[a]			
Groundnut oil	1.16	6.72	8.60
Palm oil	3.65	9.56	138.64
	Groundnut Oil Exports		
West African countries[a]	240.06	227.86	112.44
AGC countries[b]	240.50	236.15	117.67
Senegal	136.83	172.41	102.93

[a]West African countries include The Gambia, Mali, Niger, Nigeria, and Senegal.
[b]The African Groundnut Council (AGC) members include the five countries above plus Sudan.

During the period of this report, African markets continually increased their share in world imports of two major vegetable oils, palm and groundnut oils (Figure 16). At the same time, the share of AGC exporters in world exports of groundnut oil fell by two-thirds (Figure 17). As in overseas markets, most of the increase in import demand on African markets has been captured by competing vegetable oils such as palm oil. However, the shift in the relative share of individual oil products shows noticeable differences among subregional markets. As can be seen from Figures 18 and 19, import demand for groundnut oil has grown much faster in the West African submarket, whereas the bulk of the increase in palm oil import demand went to other parts of the continent. As a result, the ratio of palm oil to groundnut oil imports for the West African submarket is about 16 (close to the world average of 12), compared with 32 for Africa as a whole (Table 25). These geographic differences in the evolution of vegetable oil demand are very important, since five out of the six AGC members, as well as the main exporting country, Senegal, are all located in West Africa.

Moreover, Figures 18 and 19 show that the surge in palm oil demand is a recent phenomenon, especially in West Africa, which may have a lot to do with the poor performance of the groundnut sector in that region. Groundnut production in AGC countries declined nearly 3 percent a year between 1962 and 1987, while world production of palm oil increased at an annual rate of 8 percent during the same period (Kinteh and Badiane 1990). More importantly, the dramatic fall of groundnut exports in Nigeria, now the largest importer of vegetable oils in the region, has been a major boost to palm oil imports into the region (Figure 17).

Analyzing the factors that determine regional import demand for oilseed products as a whole and for groundnuts from AGC and non-AGC countries provides insight into the possible role of regional markets as future destinations for groundnut exports

Figure 16—Share of Africa in world imports of palm oil and groundnut oil, 1963-87

Percent

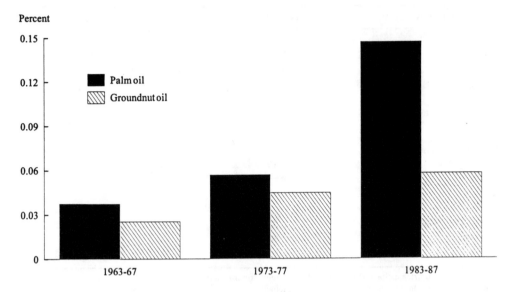

Source: Food and Agriculture Organization of the United Nations, *FAO Trade Yearbook*, (Rome: FAO, various years).

Figure 17—Export share of selected AGC countries in groundnut oil, 1961-87

Percent

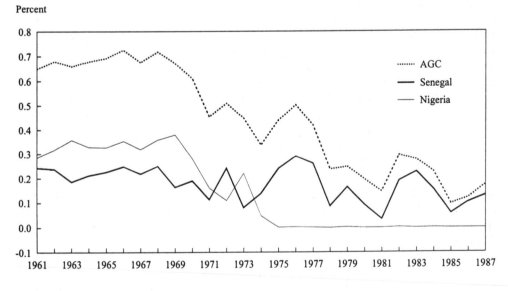

Source: Food and Agriculture Organization of the United Nations, *FAO Trade Yearbook*, (Rome: FAO, various years).
Note: AGC is African Groundnut Council.

Figure 18—Changes in groundnut oil imports between 1963 and 1987

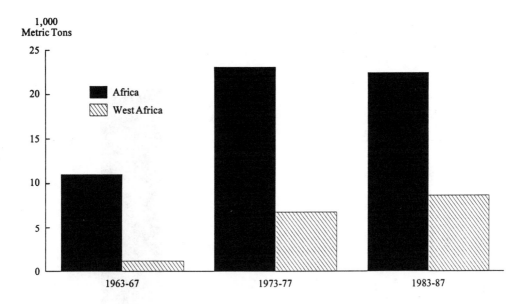

Source: Food and Agriculture Organization of the United Nations, *FAO Trade Yearbook*, (Rome: FAO, various years).

Note: West Africa includes the main importers only: Côte d'Ivoire, Ghana, and Nigeria. Africa includes West Africa.

by AGC countries.[20] Hence the analysis in this section focuses primarily on the importance of price competitiveness for recapturing regional markets, as indicated by the effect of relative import prices on the level of demand for individual oilseed imports, and the potential for future growth in regional markets for oilseeds, as the economies of the region recover from the economic stagnation of the last decade. This recovery is based on the income responsiveness of demand for individual oilseed imports.

The analysis of regional import demand is based on an econometric model that follows a two-stage budgeting approach, which distinguishes, first, between demand for groundnut imports and that for other oilseeds, and, second, between groundnut imports from AGC and non-AGC sources. The distinction based on import sources implies that product differention is assumed, meaning that regional traders and consumers do not treat groundnut products from the two sources as perfect substitutes. Despite physical similarities, there are a number of factors that support the assumption of product differentiation between groundnut imports from AGC and non-AGC sources. Most important is the historical bias in the infrastructure for import and distribution in favor

---

[20]Ideally, forecasting future demand levels would be part of the analysis of the importance of regional markets for future AGC exports. The analysis presented here is restricted to investigating the reaction of regional demand for oilseed products to key determining variables in order to obtain some idea, first, of the possible evolution in groundnut import demand and, second, factors AGC countries will have to pay attention to if they want to take advantage of any expansion in regional demand.

Figure 19—Changes in palm oil imports between 1963 and 1987

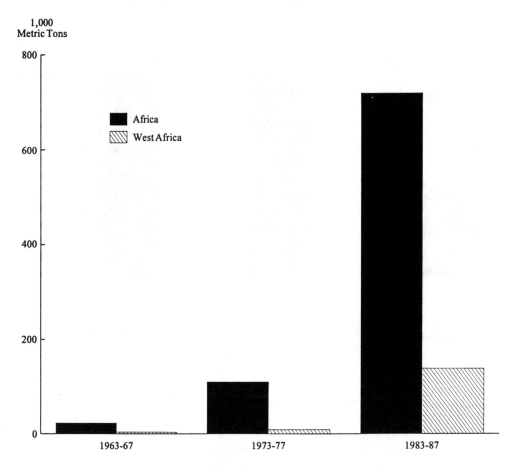

Source: Food and Agriculture Organization of the United Nations, *FAO Trade Yearbook*,
          (Rome: FAO, various years).
Note: West Africa includes the main importers only: Côte d'Ivoire, Ghana, and Nigeria. Africa includes West Africa.

of extraregional imports. The control of groundnut and extraregional oilseed trade channels by a limited number of foreign-based corporations and state-trading organizations also reinforces the bias and provides the opportunity to discriminate between trade flows. Similarly, factors related to packaging, reliability, and timeliness of delivery may lead to different behavior on the part of traders and consumers regarding imports from the two sources.

Furthermore, the physical similarity of raw material ceases in many cases once they are processed. AGC groundnuts are mainly exported in the form of oil. As for many other processed products, the place of origin is an important parameter that influences the decision of local consumers. Many local manufacturers react to this behavior by leaving out information relating to the origin of their products or by deliberately putting a wrong but more acceptable label on them. A good illustration is the decision by Senegal's main mineral water bottling company to adopt a label showing snowy mountains and a European family in order to increase the acceptance of its product on domestic and regional markets.

In accordance with the assumption of product differentiation and following Badiane (1988) and Honma (1991), a model based on two-stage maximization behavior is used to analyze regional demand for oilseed imports.[21] Applying product differentiation to oilseed imports means that two product categories can be distinguished for each individual type of oilseed. The first category corresponds to each variety of oilseed differentiated by its source of origin, and the second corresponds to aggregates of similar oilseeds from different sources. Using groundnuts as an example, the first category designates whether groundnuts are from AGC or non-AGC sources, while the second category refers to groundnuts in general, without any source distinction.

Given functional separability, consumers' overall utility can, therefore, be viewed as a function of subutilities, following the staged structure of maximization.[22] At the first stage, consumers seek to maximize top-level utility (equation 35) in demanding optimum quantities of the composite goods (equation 37), based on available income (equation 36) and the price index of each commodity aggregate. That is,

$$U = U(Q'''), \tag{35}$$

where U is utility and Q''' is the quantity of imports.

$$E = \Sigma_i Q_i''' P_i''', \tag{36}$$

where E is expenditure and Q_i''' and P_i''' are defined below.

$$Q_i''' = Q_i'''(P''', E), \text{ and} \tag{37}$$

$$P_i''' = P_i'''(P_{is}'''). \tag{38}$$

In these equations as well as the following ones, subscript i denotes commodities, such as groundnuts, and subscript s is the source of regional imports, that is AGC and non-AGC sources. Q_i''' is the quantity index of oilseed imports at the top level, that is not differentiated by source. P''' is the vector of price indices for individual oilseed imports at the same level. Its elements, P_i''', are functions of the prices of imports from AGC and non-AGC countries, P_{is}''', as indicated in equation (38).

Given the utility maximizing quantities (equation 37) of individual commodity categories, which represent aggregates of imports from AGC and non-AGC countries (equation 40), consumers then choose the levels of imports from the two sources (equation 41) that minimize expenditure in each category, as given in equation (39).

---

[21]For a discussion of the model and other examples of its application, see Badiane 1988, Chapter 5, and Honma 1991, Chapter 3.

[22]For discussion of functional separability and its implications for staged optimization, see Varian 1984 and Shiells, Stern, and Deardorff 1986.

$$E_i^m = \Sigma_s Q_{is}^m P_{is}^m, \tag{39}$$

$$Q_i^m = Q_i^m(Q_{is}^m), \text{ and} \tag{40}$$

$$Q_{is}^m = Q_{is}^m (P_{i.}^m, Q_i^m), \tag{41}$$

where Q_{is}^m denotes individual oilseed imports from AGC and non-AGC sources, $P_{i.}^m$ is the vector of corresponding prices, and E_i^m is the combined expenditure on imports of single oilseeds from the two sources.

Equations (37) and (40) can now be used for the econometric analysis of regional oilseed imports. For that purpose the two equations are expressed in double-logarithmic form and the error terms are added to yield estimating equations for regional import demand for groundnuts and competing oilseeds (equation 42) and for groundnut products from AGC and non-AGC countries (equation 43).

Thus,

$$\ln Q_i^m = \beta_i + \Sigma_j \beta_{ij} \ln P_j^m + \beta_{ie} \ln E + u_i, \tag{42}$$

where Q_i^m denotes the quantity indices of imports of individual oilseeds, and β_{ij} and β_{ie} are, respectively, the own and cross-price elasticities and the income elasticity of import demand for oilseed i. The import prices for individual oilseed varieties, P_j^m as well as total income, E, are deflated using the consumer price index in the importing country.

$$\ln Q_s^m = \lambda_s + \Sigma_f \lambda_{sf} \ln P_f^m + \lambda_{se} \ln E^m + v_s, \tag{43}$$

where Q_s^m stands for regional demand for oilseed imports from AGC and non-AGC countries, and P_f^m and E^m are, respectively, the unit value of imports of individual oilseeds from the two sources and the combined expenditure on the same imports. They are all deflated using the price index for total imports of the corresponding oilseed, given in equation (38). The coefficient of the expenditure variable, λ_{se}, expresses the responsiveness of import demand for oilseed i from AGC and non-AGC sources with respect to total expenditure on imports of the same oilseed. The coefficients of the price variables, λ_{sf}, are the conditional import price elasticities, which indicate the responsiveness of import flows from the two sources with respect to their own prices, given the level of overall import demand for the same oilseed. Using the two sets of parameters and the coefficients in equations (42), the own price (δ_{ss}) and income (ϕ_{is}) elasticities, which give the overall reaction of AGC and non-AGC import flows to changes in income and import prices, can be calculated as in expressions (44) and (45), respectively (Honma 1991, 40-41):

$$\delta_{ss} = \lambda_{ss} - CM_s (\Sigma_f \lambda_{sf} - \lambda_{se} \beta_{ii}), \tag{44}$$

---

[23] As already indicated in equation (38), prices of composite imports are based on prices for imports from the different sources. For the estimations, $P_{i.}^m$ is calculated as a weighted index of the unit value of imports from the different sources, P_{is}^m using the share of individual sources in the imports of each oilseed as weights.

where CM_s denotes the share of imports from source s in total imports of commodity i, and

$$\phi_{is} = \lambda_{se} \beta_{ie}. \tag{45}$$

The first-stage demand equation (equation 42) is estimated separately for the three main oilseeds imported by West African countries—groundnuts, oil palm, and soybeans.[24] As indicated in Table 9, regional imports of these oilseeds consist mainly of oils. The estimations are, therefore, based on data for oil imports only. Data from the two main importers of vegetable oils in the region, Côte d'Ivoire (for groundnut oil) and Nigeria (for palm and soybean oils) are used in the estimations.[25]

The second-stage equation, (equation 43), which represents demand for imports from AGC and non-AGC countries, is estimated only for groundnut oil, given the focus of the study on groundnut exports by the AGC countries to regional markets. Data on regional groundnut oil imports from Senegal, the main exporting AGC-member country, are used in the estimation. Given the unavailability of data on Nigerian imports of groundnut oil from AGC countries, for example, Senegal, only import flows to Côte d'Ivoire are used.

Unlike equation (42), the error terms in equation (43) are expected to be correlated contemporaneously, because of disturbances that are not captured by variables included in the model but that affect overall demand for groundnut imports and, therefore, affect imports from AGC and non-AGC countries alike. Hence, equation (43) is estimated based on a modification of Zellner's seemingly unrelated regressions (SUR) technique to account for unequal numbers of observations across equations due to shorter time series for non-AGC import flows (see Schmidt 1977; Judge et al. 1985).

Furthermore, in estimating overall regional demand for individual oilseed imports, dummy variables are introduced to capture changes in the structure of demand for these products. The oil boom of the 1970s in Nigeria, the dramatic fall in oilseed production in that country, and the severe drought at the end of the 1960s and early 1970s in Senegal are expected to have affected the structure of regional demand for vegetable oils. The cusum-squared test is used to identify periods of structural change in regional demand and dummies included for the corresponding years to capture its effect.

The estimations with the best fit are presented in equations (46) to (51).[26] The estimates for the regional import demand parameters and the derived own-price and income elasticities for AGC and non-AGC groundnut imports are summarized in Tables 26 and 27. The very high estimates of own-price and income elasticities in Table 26 indicate that regional imports of the three oilseeds in the table are very sensitive to changes in import prices and incomes. The income elasticities are

---

[24]The error terms of the individual oilseed demand equations could be expected to be contemporaneously correlated. However, the chi-square test for the palm and soybean import equations, which are based on Nigerian import data, failed to reject the hypothesis of a diagonal covariance matrix. The groundnut import equation is based on Côte d'Ivoire data alone and is, therefore, not included in the test.

[25]The estimates do not include data from Ghana because regional imports of palm and soybean oils are heavily dominated by Nigeria, and estimations for groundnuts based on Ghanaian data yield insignificant coefficients.

[26]The data used for the estimations are presented in a supplement to this report, available on request from the International Food Policy Research Institute.

Table 26— Estimates of demand parameters for regional oilseed imports

Imported Oilseed	Own Price (β_{ii})	Income Elasticity (β_{ie})
Groundnuts[a]	−1.77	2.76
Palm oil[b]	−4.07	2.33
Soybeans[b]	−2.29	2.85

Sources: Import quantities and price data are from Food and Agriculture Organization of the United Nations, *FAO Trade Yearbook* (Rome: FAO, various years). Income data are from World Bank, *World Tables, 1989-90*, Baltimore, Md., U.S.A.: Johns Hopkins University Press, 1990.

Notes: Estimates are based on data for vegetable oils, which are the main oilseed products imported by West African countries. All estimates are significant at the 0.05 level.

[a]Based on Côte d'Ivoire import data for the period 1966-85.

[b]Based on Nigerian import data for the period 1976-87.

comparably high for all three, suggesting a rapid increase in regional demand, as the economies of the region expand.

In contrast, the price elasticities show strong differences across oilseeds. Palm oil and, to a lesser extent, soybeans display a much higher degree of responsiveness to import price changes than do groundnuts. This means that equal advances in cutting unit costs of production and distribution translate into much higher gains in regional demand for these two products than for groundnuts. Consequently, it appears that groundnut exporters, in general, would face increased competition on regional markets, if palm oil- and soybean-producing countries keep or expand their technological advantages of the past.[27]

Table 27— Estimates of demand parameters for groundnut imports from AGC and non-AGC sources

Origin of Imports	Coefficients of Import Equation		Import Demand Elasticities	
	Own-Price Coefficient (λ_{ss})	Expenditure Coefficient (λ_{se})	Own-Price Elasticity (δ_{ss})	Income Elasticity (δ_{is})
AGC[a]	−3.37	0.84[b]	−2.17	2.32
Non-AGC[a]	−1.17	1.21[b]	−2.70	3.34

Notes: All coefficients are significant at the 0.005 level. Import demand elasticities are computed using equations (44) and (45) and import shares of 0.64 for African Groundnut Council (AGC) and 0.36 for non-AGC countries. Shares do not add up to 1.00 because estimates for the two sources are carried out for different time periods.

[a]Based on groundnut exports from Senegal (AGC) and the rest of the world (non-AGC) to Côte d'Ivoire, the only groundnut-importing country in the region for which data on import flows by sources are available.

[b]Tests of the homotheticity hypothesis confirm the divergency of each expenditure coefficient from unity.

---

[27]The significantly higher and primarily yield-driven production growth rates for oil palm and soybeans in Table 1 (column 2) are a good illustration of the technological edge enjoyed by countries producing these two commodities, compared with African (mainly AGC) producers.

Together with the figures in Table 26, the estimates for the second-stage import equations in Table 27 show that groundnut imports are sensitive to prices in terms of both aggregate imports and imports from the two different sources. However, demand for groundnut imports by sources (third column of Table 27) seems to be more responsive to price changes than overall groundnut import demand (first column of Table 26). The higher price responsiveness of import demand at the second stage reflects the process of substitution between AGC and non-AGC exports that accompanies the adjustment in regional demand for groundnut imports. Moreover, the significantly higher value of the own-price coefficient for AGC groundnuts in Table 27 (–3.38) suggests that, by cutting their costs of production and distribution, AGC exporters could increase both the quantity and value of their exports to regional markets and raise their market share.

Demand for individual oilseeds is expressed in equations (46) to (48).

Groundnuts:

$$\ln Q_{GN}^m = -24.15 - 1.77 \ln P_{GN}^m + 1.02 \ln P_{OL}^m + 2.76 \ln E \; ; \tag{46}$$
$$\phantom{\ln Q_{GN}^m = } (3.33) \; (2.22) \phantom{\ln P_{GN}^m} (3.02) \phantom{\ln P_{OL}^m} (3.04)$$

MA1: Box-Pierce statistic = 9.24; degrees of freedom = 3.

This equation is based on equation (42) applied to groundnut oil imports by Côte d'Ivoire for the period 1969-87.

Palm oil:

$$\ln Q_{PL}^m = 0.73 - 4.07 \ln P_{PL}^m + 7.36 D \times \ln P_{PL}^m - 3.70 \ln E \; ; \tag{47}$$
$$\phantom{\ln Q_{PL}^m = } (0.26) \; (2.81) \phantom{\ln P_{PL}^m} (2.37) \phantom{\ln P_{PL}^m} (2.33)$$

AR1: $\overline{R}^2 = 0.80$; DW = 1.82.

This equation is based on equation (42) applied to palm oil imports by Nigeria for the period 1976-87.

Soybeans:

$$\ln Q_{SY}^m = -11.66 - 2.29 \ln P_{SY}^m + 2.85 \ln E \; ; \tag{48}$$
$$\phantom{\ln Q_{SY}^m = } (6.20) \; (2.81) \phantom{\ln P_{SY}^m} (2.45)$$

OLS: $\overline{R}^2 = 0.49$; DW = 1.90.

This equation is based on equation (42) applied to soybean oil imports by Nigeria for the period 1976-87.

Demand for AGC and non-AGC groundnuts is expressed in equations (49) to (51). The three groundnut equations are based on equation (43) applied to groundnut oil imports by Côte d'Ivoire for the period 1966-85. For AGC groundnuts, both SUR and OLS estimates are presented.

AGC groundnuts:

$$\ln Q_{AGC}^m = -0.81 - 3.38 \ln P_{AGC}^m - 0.81 \ln P_{ROW}^m + 0.84 \ln E^m \; ; \tag{49}$$
$$\phantom{\ln Q_{AGC}^m = } (65.22)(16.06) \phantom{\ln P_{AGC}^m} (13.05) \phantom{\ln P_{ROW}^m} (679.60)$$

SUR: $R^2 = 0.94$; F = 83.56.

$$\ln Q^m_{AGC} = -0.80 - 3.36 \ln P^m_{AGC} - 0.81 \ln P^m_{ROW} + 0.84 \ln E^m \; ; \qquad (50)$$
$$\quad\;\; (4.50) \;\; (4.65) \qquad\quad (2.06) \qquad\quad (15.06)$$

OLS: $\overline{R}^2 = 0.92$; DW = 2.13.

Rest-of-world groundnuts:

$$\ln Q^m_{ROW} = -0.80 + 2.63 \ln P^m_{AGC} - 1.17 \ln P^m_{ROW} + 1.21 \ln E^m \; ; \qquad (51)$$
$$\quad\;\; (15.97) \;\; (3.34) \qquad\quad (3.96) \qquad\quad (136.61)$$

SUR: $R^2 = 0.83$; F = 13.02.

Furthermore, the estimates yield an expenditure elasticity value that is almost 50 percent higher for non-AGC than for AGC groundnut imports (Table 27, column 2). Testing the individual values of the expenditure elasticity against unity (the homotheticity test) indicates the extent to which AGC and non-AGC exporters may benefit from expanding regional markets for groundnut products. For that purpose, chi-square tests are carried out against the hypothesis $H_0 : \lambda_{se} \geq 1$ for AGC exports and against the hypothesis $H_0 : \lambda_{se} \leq 1$ for non-AGC exports. In both cases, the H_0 hypothesis is rejected, meaning that AGC exports would expand less rapidly and non-AGC exports more rapidly than aggregate regional demand for groundnut imports. Similarly, the higher income elasticity coefficient for non-AGC groundnut imports (Table 27, column 4) indicates that non-AGC suppliers would profit more from increases in regional incomes than would AGC suppliers, suggesting that the structure of regional import demand favors the former over the latter.

Based on these results and the ones from the constant market share model, AGC exporters do not seem to enjoy any real advantage from their proximity to regional markets. The findings put to question the success of strategies to revitalize national groundnut sectors by encouraging regional outlets as alternatives to traditional export markets. The expected fast-growing future demand on regional markets (Table 23), the high price and income elasticities of groundnut import demand, and an expenditure elasticity for groundnut import demand from AGC countries that is not much below unity all indicate that regional markets could play an important role in future AGC exports.

7

CONCLUSIONS

Although it is still significant in some cases, the contribution of the groundnut sector to the economies of AGC countries has diminished consistently since the 1960s. Both production and exports fell sharply during the 1970s and 1980s. These dramatic changes in the groundnut economies have often been portrayed by country policymakers as a result of shrinking demand on international markets.

The data presented in this report show that the decline in AGC exports between the 1960s and the 1980s was about three times larger than the fall in global groundnut oil exports, which was about 45,000 metric tons. Consequently, the decline in AGC export performance can hardly be explained by reduced external demand. What the data also show is that global demand for vegetable oils tilted away from groundnuts toward competing oils, the exports of which grew two times faster. Although this tendency may raise some doubts about long-term prospects for groundnut oil relative to other vegetable oils, it should not have much impact on the performance of individual exporters on world groundnut markets.

Moreover, the share of AGC countries in world groundnut exports has fallen by more than 50 percent during the period 1961-65 to 1986-88, while exporters from South American and Asian countries have more than quadrupled their combined share. AGC's loss of market share was also accompanied by a continuous decrease in yield and acreage in its member countries, contrasted with strong yield increases in competing Asian and South American countries. It would seem, therefore, that AGC exports have suffered more from domestic than external demand factors.

In fact, the role of country policies in reducing incentives to groundnut production and trade appear overwhelming. The net level of policy-induced implicit taxation of the groundnut sector ranged between 10 and 30 percent in The Gambia, Senegal, and Sudan. In The Gambia and Senegal, the level of direct and indirect disincentives facing groundnut production and trade was particularly high during the 1970s, a period when the world groundnut economy was booming, but also the period when AGC countries suffered the greatest losses in market shares.

Due to the strong suppression of prices and incentives in the groundnut sector, domestic policies had a detrimental impact on groundnut production and exports. Except for Sudan for a few years and for the last few years of the study, domestic policies kept country production and exports well below the levels that would have prevailed. At the prevailing export prices, the reduction in output and export quantities caused yearly export revenue losses of 20-70 percent for Senegal and The Gambia. For Sudan, the computed revenue losses varied from 10 to 20 percent. Translated into changes in aggregate AGC groundnut exports, these losses correspond to a decline of up to 31 percent in quantity and 54 percent in annual export revenue.

These results clearly indicate that domestic policies contributed significantly to the decline of the groundnut sector in AGC countries and to a significantly larger extent than factors related to international groundnut and oilseed markets. These policies had strong detrimental effects, both directly and indirectly, on prices and incentives. They suppressed producer prices directly and caused country real exchange rates to appreciate significantly. The ultimate consequence was a substantial reduction in output, export volumes, and export revenues in individual member countries and for the AGC as a whole.

The primary role of domestic factors in the decline of the groundnut sector in AGC countries is also supported by the findings of another AGC-mandated study, which found that deficiencies in the production of and access to improved seeds and in the availability and distribution of fertilizer have had a significant and negative impact on the performance of the AGC's groundnut sector (UNECA/FAO 1985).

In the debates among AGC officials and country policymakers, the idea of recapturing regional vegetable oil markets to compensate for dwindling international outlets has received increased attention. Based on the results obtained from the constant market share model, however, AGC countries have not taken advantage of the proximity of regional markets in the past. Although demand in regional markets— other African countries—grew two-and-a-half times faster than world markets, the AGC countries seldom exported to regional markets.

Moreover, estimates of own-price and income elasticities reveal a high degree of sensitivity of demand for individual oilseed imports with respect to changes in import prices and incomes. The high income elasticities for groundnuts and soybeans especially suggest a rapid increase in regional demand for them, as the economies of the region expand. In contrast, the estimated income elasticity for palm oil is negative, indicating that it is viewed as an inferior good and that it will become less of a competitor to groundnuts as the economies of the region grow and incomes increase.

The estimates for price elasticities indicate a much higher degree of price-responsiveness for palm oil and, to a lesser extent, soybean imports than groundnuts. This means that equal advances in cutting unit costs of production and distribution translate into much higher gains in regional demand for these two products than for groundnuts. It appears that groundnut exporters will face increased competition on regional markets if palm oil- and soybean-producing countries keep or expand their past technological advantages.

Furthermore, the analysis of demand for groundnut imports by sources suggests regional importers and consumers adjust to price changes by substituting between imports from AGC and non-AGC sources. The estimates obtained for the own-price elasticity for AGC and non-AGC groundnut imports suggest that, even with stagnant demand, AGC exporters could increase both the quantity and value of their exports to regional markets and raise their market share by cutting their costs of production and distribution.

In contrast, the estimates of the expenditure elasticities yield a value that is one-and-a-half times higher for non-AGC than for AGC groundnut imports, an indication that AGC exports would expand less and non-AGC exports more rapidly than aggregate regional demand for groundnut imports. Similarly, the higher income elasticity coefficient obtained for non-AGC groundnut imports means that non-AGC suppliers would profit more from increases in regional incomes than would AGC suppliers, suggesting that the structure of regional import demand favors the former over the latter.

The finding that AGC exporters do not take much advantage of their proximity to regional markets raises the question of whether strategies to revitalize national groundnut sectors through increased exports to regional outlets is realistic. However, demand on regional markets is likely to grow rapidly, as indicated by high price and income elasticities of groundnut import demand and a relatively high import expenditure elasticity for AGC groundnuts. All of these indicators mean that regional markets could play an important role in the future of AGC exports, but only if member countries cut costs in production, marketing, and other export-related activities in order to contain the competition from non-AGC exporters.

This stresses the vital need to adjust the domestic policy environment in AGC countries in order to eliminate the detrimental effects of sector and macroeconomic policies on prices and incentives in their groundnut sectors. The liberalization of domestic markets and the other reforms pertaining to marketing that have been initiated in most AGC member countries are an important step toward eliminating the disprotection of country groundnut sectors evidenced in this study and toward restoring their competitiveness.

The same applies to reforms of macroeconomic policies and overall trading regimes. In light of the significant impact they have on incentives in the groundnut sectors, it is doubtful that AGC countries could even take advantage of favorable demand conditions on international or regional export markets without effective changes in the domestic policy environment.

The hope expressed at the Banjul meeting that intensifying regional trade in groundnut products will help solve the problems faced by AGC exporters will not be realized unless appropriate changes in domestic policies that affect production and trade in the groundnut sector take place. AGC exports to the region and elsewhere have suffered much more from these domestic policies than from external demand constraints. However, the debates on increased regional trade and integration can help groundnut exports to regional markets if they contribute to lowering the cost of moving products through local and transborder markets.

APPENDIX 1: FINDINGS AND RECOMMENDATIONS OF A JOINT STUDY BY THE UNITED NATIONS ECONOMIC COMMISSION FOR AFRICA AND THE FOOD AND AGRICULTURE ORGANIZATION OF THE UNITED NATIONS

National Seed Production Programme

One major reason for the decline in groundnut production in the AGC-member countries is the fact that most farmers lack improved high quality seeds which can give them relatively high yield. The shortage of improved seeds is widespread throughout the AGC-member States and this shortage has contributed immensely to the prevailing low yield obtained for groundnut in each of the member States.

Given the fact that the most important input for rehabilitation of groundnut is the production and supply of improved seeds, the need for seed production programme cannot be over emphasized. Although considerable research has been done on the development of improved seeds, most farmers in the AGC-member countries have not benefitted significantly from this research due to lack of seed farms to supply farmers with high quality seeds.

As a major step towards groundnut rehabilitation, each country should set up seed production farms to be located in various ecological zones in their groundnut producing areas. Since Senegal is already advanced in seed development, the help of AGC in obtaining some information on high quality seeds from Senegal can be sought. Also, since the establishment of seed farms would involve some financing, each country can seek the help of FAO and UNDP in financing part of the seed farm project. The FAO technical cooperation programme can assist each country not only with the partial financing of the project but also with the technical guidance and management of the seed farms. The AGC through their Technical and Scientific Department can also participate in the seed programme by providing technical information and expertise to backup the seed programme.

In summary each country should embark on [a] seed farm project to be financed jointly by their government and possibly by FAO/UNDP. The AGC should participate in the project by providing some technical support and assistance.

Subsidy on Improved Seeds

The improved seeds produced under the seed project should be subsidized by each State government as a means of encouraging farmers to purchase and use them for their plantings. Given the fact that most of the groundnut farmers are smallholders having very low income, most of the farmers may be unable to purchase and use the improved seeds if the prices are not subsidized. The recommended level of subsidy

This appendix is excerpted from a report on a study carried out by the United Nations Economic Commission for Africa and FAO's Agriculture Division, Addis Ababa (see UNECA/FAO 1985, 2).

should not be lower than 50 percent. This is the only way of ensuring that the improved seeds produced are widely adopted for plantings by the groundnut farmers.

Fertilizer Programme

Lack of fertilizer used by the groundnut farmers has been one of the factors responsible for the present low level of yield obtained in AGC-member States. To ameliorate this situation it is recommended that each member country should establish a National Fertilizer Programme whereby the required quantities of fertilizer would be imported into the country and made available at the right time and at a subsidized rate to the groundnut farmers.

Since shortage of foreign exchange has been a major constraint in the importation of fertilizers each country can seek the assistance of FAO and the World Bank with respect to financing the fertilizer programme. Also, bilateral loans and grants could be sought between each member country in connection with the proposed fertilizer programme. The Gambian government already has a grant from Italy for its fertilizer project although the grant is still inadequate to meet her total requirements. Where grants are not available direct loans can be obtained from the World Bank to finance the proposed fertilizer programme.

Village Industrial Process in Programme

Given the present groundnut processing constraints at the village levels in most of the AGC-member States, one way of rehabilitating the declining groundnut industry is to establish small-scale groundnut processing units at the village level. Such village processing units can follow the Indian pattern and each member country can seek the assistance of UNDP/UNIDO in setting-up the proposed project. Also the consultancy advice of UNECA/UNIDO Division can be sought by AGC in setting up the individual units in the various member countries.

UNIDO and each member country would be expected to collaborate in financing and managing the project.

APPENDIX 2: ESTIMATION OF THE EQUILIBRIUM EXCHANGE RATE

Restrictive trade regimes and imbalances in overall economic policies are typically reflected in a sustained appreciation of the real exchange rate and a deterioration of country trade balances. Accordingly, a model linking the exchange rate to trade restrictions and the current account deficit is used to estimate the equilibrium exchange rate(E^e).[28] It is assumed that individual country supply X_s and demand (M_d) for foreign exchange react to changes in the real exchange rate (E) with elasticities ε_s and η_d respectively defined as

$$\varepsilon_s = (dX_s/X_s)/(dE/E), \text{ and} \tag{52}$$

$$\eta_d = (dM_d/M_d)/(dE/E). \tag{53}$$

First, defining E^a as the actual official exchange rate and X_a and M_a as the actual levels of aggregate country exports and imports; second, defining Q^t as the equilibrium level of exports and imports in the absence of trade restrictions; and third, defining E^t as the value of the balanced-account exchange rate, ε_s and η_d can be rewritten as

$$\varepsilon_s = [(Q^t - X_a)/X_a]/[(E^t - E^a)/E^a], \text{ and} \tag{54}$$

$$\eta_d = [(M_a - Q^t)/M_a]/(E^t - E^a)/E^a]. \tag{55}$$

Furthermore, if the unsustainable part of the balance of trade (B_a) is defined as $B_a = M_a - X_a$, equations (54) and (55) can be solved to yield[29]

$$B_a = [(E^t - E^a)/E^a](\varepsilon_s X_s + \eta_d M_d), \text{ and} \tag{56}$$

$$(E^t - E^a)/E^a = B_a/(\varepsilon_s X_s + \eta_d M_d). \tag{57}$$

Equation (57) gives the change in the exchange rate that is required to eliminate the unsustainable part of country current account deficits.

Since country exchange rates are equally affected by the imposition of trade restrictions, equation (57) needs to be modified to include the change in the exchange rate that would arise from the removal of trade restrictions. In the presence of

---

[28]The model is based on Krueger, Schiff, and Valdés 1988. For other applications, see Stryker 1990; Intal and Power 1990; Jansen 1988; Jenkins and Lai 1989; Moon and Kang 1989; Garcia and Llamas 1989.

[29]Specification of the sustainable level of country current account balance rests on assumptions about the normal level of financial flows. Given the difficulty this presents in identifying the sustainable share of actual country current account imbalance, the calculations carried out in the study are based on sustainable levels of zero current account balance.

restrictions, the true exchange rates received by exporters (E_s^t) or paid by importers (E_d^t) in each of three countries—The Gambia, Senegal, and Sudan—differ from E^t, the corresponding country's equilibrium exchange rate. The former are determined by the actual equivalent rates of taxation of exports (t_x) and imports (t_m) in each country, as presented in expressions (58) and (59):

$$E_s^t = (1 - t_x) E^t, \text{ and} \tag{58}$$

$$E_d^t = (1 + t_m) E^t. \tag{59}$$

The effect of trade restrictions on the current account can thus be calculated as

$$B_t = \eta_d [(E_d^t - E^t)/E^t] M_d - \varepsilon_s [(E_s^t - E^t)/E^t] X_s, \tag{60}$$

with the effect of removing trade restrictions on country import and export prices given by

$$(E_d^t - E^t)/E^t = t_m/(1 + t_m), \text{ and} \tag{61}$$

$$(E_s^t - E^t)/E^t = t_x/(1 - t_x). \tag{62}$$

Inserting equations (61) and (62) into equation (60) yields a new expression for the impact of trade restrictions on the balance of trade:

$$B_t = \eta_d [t_m/(1 + t_m)] M_d - \varepsilon_s [t_x/(1 - t_x)] X_s. \tag{63}$$

Adding B_t as defined in equation (63) to B_a in equation (57) yields the change in exchange rates that would prevail in a situation without trade restrictions and with balanced country current accounts. The new expression is

$$(E^e - E^a)/E^a = (B_a + B_t)/(\varepsilon_s X_s + \eta_d M_d). \tag{64}$$

Equation (64) can now be solved for the equilibrium exchange rate E^e, which would prevail in the absence of trade restrictions and other domestic policies that cause country exchange rates to appreciate. The expression for E^e, is

$$E^e = [(B_a + B_t)/(\varepsilon_s X_s + \eta_d M_d) + 1] E^a. \tag{65}$$

Or, using the expression for B_t in equation (63),

$$E^e = E^a \frac{B_a + \eta_d [t_m/(1 + t_m)] M_d + \varepsilon_s [t_x/(1 - t_x)] X_s}{\varepsilon_s X_s + \mu_d M_d} + E^a. \tag{66}$$

APPENDIX 3: THE CONSTANT MARKET SHARE MODEL

The model adopted for this report is similar to the one developed in Magee 1975. It starts with the following identity:

$$S_{t1} = R \cdot S_{t0}, \tag{67}$$

where S_{t0} denotes the shares of a given member country in total world exports of oilseeds in the beginning period (1962-67) and S_{t1} in the end period (1982-87). R represents a relative growth factor defined as follows:

$$R = \frac{(1 + g^m)}{(1 + g^w)}, \tag{68}$$

where g^m and g^w stand for the percentage growth rates of total exports of oilseed and oilfruit products of country m and the world w between the beginning and the end period. Equation (68) expresses the growth of country m's exports (X^m) relative to world exports and can be rewritten as

$$R = \sum_i \frac{(1 + g_i^m)}{(1 + g^w)} \frac{(X_{it0}^m)}{(X_{t0}^m)}, \tag{69}$$

with $X_t^m = \sum_i X_{it0}^m$. By expressing X for the different products i and different export destinations j in equation (69), multiplying by $[(1 + g_i^w)/(1 + g_i^w)]$ and by $[(1 + g^w)X_{t0}/(1 + g^w)X_{t0}^m]$, summing over i and j, and rearranging the terms, the following result is obtained.

$$R = \sum_i \frac{(1 + g_i^m)}{(1 + g_i^w)} \frac{(1 + g_i^w)}{(1 + g^w)} \frac{X_{it0}^m}{X_{t0}^m} \sum_j \frac{(1 + g_j^w)}{(1 + g^w)} \frac{X_{jt0}^m}{X_{t0}^m}, \tag{70}$$

where $X_{t0}^m = \sum_i X_{it0}^m = \sum_j X_{it0}^m$, and i and j represent individual oilseed and oilfruit products and export destinations, respectively. By substituting equation (70) for R in equation (67), the result is a new expression for the change in country export shares between 1962-67 and 1982-87:

$$S_{t1} = S_{t0} \underbrace{\sum_i \frac{(1 + g_i^m)}{(1 + g_i^w)}}_{a} \underbrace{\frac{(1 + g_i^w)}{(1 + g^w)} \frac{X_{it0}^m}{X_{t0}^m}}_{b} \underbrace{\sum_j \frac{(1 + g_j^w)}{(1 + g^w)} \frac{X_{jt0}^m}{X_{t0}^m}}_{c}. \tag{71}$$

It is clear from equation (67) that the direction of a country's export share during a given time period depends on whether the relative growth factor R is greater than, less than, or equal to unity. Furthermore, the new expression for R in equation (71) shows that a country may increase its global trade share for several reasons: (1) it has been able to raise its exports in single-product markets faster than the world average (term a of equation [71]); (2) its exports are concentrated on the commodities that experience faster growth rates than the aggregate of oilseed and oilfruit products (the last two terms of the first sum); and (3) its exports are directed more toward markets that grow faster than the world average (term c of equation [71]).

REFERENCES

Abdoulaye, M. 1984. Problems of groundnuts production in Niger and attempts at combating them. *Proceedings of the international symposium on production, world oilseeds market and intra-African trade in groundnuts and products*, 6-11 June 1984, Banjul, The Gambia. Lagos, Nigeria: African Groundnut Council.

AGC (African Groundnut Council). 1984. *Proceedings of the international symposium on production, world oilseeds market and intra-African trade in groundnuts and products*, 6-11 June 1984, Banjul, The Gambia. Lagos, Nigeria: African Groundnut Council.

Agbola, S. D., and J. S. Opadokun. 1984. A review of groundnut quality and storage in Nigeria. *Proceedings of the international symposium on production, world oilseeds market and intra-African trade in groundnuts and products*, 6-11 June 1984, Banjul, The Gambia. Lagos, Nigeria: African Groundnut Council.

Andies, M. 1987. *The new common market in oils and fats*. Brussels: Le Club de Bruxelles.

Badiane, O. 1988. *National food security and regional integration in West Africa*. Kiel, Germany: Vauk Publishers.

Badiane, O., and S. Kinteh. 1992. Agricultural trade pessimism in West African countries: The possible role of regional markets. International Food Policy Research Institute, Washington, D.C. Mimeo.

Bond, M. E. 1983. Agricultural responses to prices in Sub-Saharan African countries. *IMF Staff Papers* 30 (No. 4): 703-726.

Drave, E. H., and Z. V. Dembele. 1984. Groundnut: Crop for international consumption or the export market? *Proceedings of the international symposium on production, world oilseeds market and intra-African trade in groundnuts and products*, 6-11 June 1984, Banjul, The Gambia. Lagos, Nigeria: African Groundnut Council.

Elbadawi, I. A. 1988. *Foreign trade, exchange rate, and macroeconomic policies and the growth prospects for Sudanese agriculture*. New Haven, Conn., U.S.A.: Yale University, Economic Growth Center.

El Bashir, A. R., and B. Idris. 1983. Impact of marketing and pricing policies on production of groundnuts. Paper prepared for a workshop on agricultural price policies in the Democratic Republic of the Sudan, sponsored by the Food and Agriculture Organization of the United Nations and the Government of Sudan, 31 May - 2 June, Khartoum, Sudan.

Elhafiz, A. T., et al. 1990. Towards alternative economic policies for Sudan. Sudan Economy Research Group Discussion Paper, University of Bremen, Bremen, Germany.

Food and Agriculture Organization of the United Nations (FAO). Various years a. *FAO production yearbook*. Rome: FAO.

_____. Various years b. *FAO trade yearbook*. Rome: FAO.

Garcia, J. G., and G. M. Llamas. 1989. *Trade, exchange rate, and agricultural pricing policies in Colombia*. World Bank Comparative Studies. Washington, D.C.: World Bank.

Gaye, M. 1991. Le retour des opérateurs privés dans la collecte des arachides: Situation après cinq années de réhabilitation des traitants. Institut Sénégalais de Recherches Agricoles, Dakar, Senegal. Mimeo.

Gulhati, R., S. Bose, and V. Atukorala. 1985. *Exchange rate policies in eastern and southern Africa, 1965-1983*. World Bank Staff Working Paper No. 270. Washington, D.C.: World Bank.

Harvey, C. 1990. *Improvements in farmer welfare in The Gambia: Groundnut price subsidies and alternatives*. Discussion Paper 277. Brighton, England: Institute of Development Studies.

Honma, M. 1991. *Growth in Japan's horticultural trade with developing countries: An economic analysis of the market*. Research Report 89. Washington, D.C.: International Food Policy Research Institute.

IMF (International Monetary Fund). Various years. *International financial statistics*. Washington, D.C.

Intal, P. S., and J. H. Power. 1990. *Trade, exchange rate, and agricultural pricing policies in the Philippines*. World Bank Comparative Studies. Washington, D.C.: World Bank.

Issaka, W. 1984. Problems of groundnut marketing in Niger. *Proceedings of the international symposium on production, world oilseeds market and intra-African trade on groundnuts and products*, 6-11 June 1984, Banjul, The Gambia. Lagos, Nigeria: African Groundnut Council.

Jammeh, S. 1987. State intervention in agricultural pricing and marketing in Senegal. Ph.D. dissertation, Johns Hopkins University, Baltimore, Md., U.S.A.

Jansen, D. 1988. *Trade, exchange rate, and agricultural pricing policies in Zambia*. World Bank Comparative Studies. Washington, D.C.: World Bank.

Jenkins, G., and A. Lai. 1989. *Trade, exchange rate, and agricultural pricing policies in Malaysia*. World Bank Comparative Studies. Washington, D.C.: World Bank.

Jones, C. W. 1986. *The domestic groundnut marketing system in The Gambia*. Cambridge, Mass., U.S.A.: Harvard Institute for International Development.

Judge, G. G., et al. 1985. *The theory and practice of econometrics*. New York: John Wiley and Sons.

Khan, M. S. 1974. Import and export demand in developing countries. *IMF Staff Papers* 21 (No. 3): 678-693.

Kinteh, S., and Badiane, O. 1990. The market potential for groundnut products facing West African countries and the possible role of regional markets. International Food Policy Research Institute, Washington, D.C. Mimeo.

Kristjansen, P., M. D. Newman, C. Christensen, and M. Abel. 1990. *Export crop competitiveness: Strategies for Sub-Saharan Africa.* Agricultural Policy Analysis Project Technical Report No. 109. Final report of the African cash crop competitiveness strategy study. Bethesda, Md., U.S.A.: Abt Associates.

Krueger, A. O., M. Schiff, and A. Valdés. 1988. Agricultural incentives in developing countries: Measuring the effect of sectoral and economywide policies. *World Bank Economic Review* 2 (No. 3): 255-272.

_____. 1992. *The political economy of agricultural pricing policy.* Vols. 1-4. Baltimore, Md., U.S.A.: Johns Hopkins University Press.

Louis Berger International. 1983. Kordofan region agricultural marketing and transport study. Report submitted to the U.S. Agency for International Development, Khartoum, Sudan. Mimeo.

Magee, S. 1975. Prices, income and foreign trade. In *International trade and finance: Frontiers for research,* ed. P. Kenen, 175-243. Cambridge, Mass., U.S.A.: Cambridge University Press.

Moon, P., and B. Kang. 1989. *Trade, exchange rate, and agricultural pricing policies in the Republic of Korea.* World Bank Comparative Studies. Washington, D.C.: World Bank.

Oyejide, T. A. 1993. Effects of trade and macroeconomic policies on African agriculture. In *The bias against agriculture: Trade and macroeconomic policies in developing countries,* ed. R. M. Bautista and A. Valdés, 241-262. San Francisco: International Center for Economic Growth and International Food Policy Research Institute, by ICS Press.

Puetz, D., and J. von Braun. 1990a. Attempts at market deregulation and institutional constraints. In *Structural adjustment, agriculture, and nutrition: Policy options in The Gambia,* ed. J. von Braun, K. Johm, S. Kinteh, and D. Puetz, 74-83. Working Papers on Commercialization of Agriculture and Nutrition 4. Washington, D.C.: International Food Policy Research Institute.

_____. 1990b. Price policy under structural adjustment: Constraints and effects. In *Structural adjustment, agriculture, and nutrition: Policy options in The Gambia,* ed. J. von Braun, K. Johm, S. Kinteh, and D. Puetz, 40-54. Working Papers on Commercialization of Agriculture and Nutrition 4. Washington, D.C.: International Food Policy Research Institute.

Quiroz, J., and A. Valdés. 1993. Agricultural incentives and international competitiveness in four African countries: Government intervention and exogenous shocks. In *Agricultural policy reforms and regional market integration in Malawi, Zambia, and Zimbabwe,* ed. A. Valdés and K. Muir-Leresche, 77-117. Washington, D.C.: International Food Policy Research Insitute.

Schmidt, P. 1977. Estimation of seemingly unrelated regressions with unequal numbers of observations. *Journal of Econometrics* 5: 365-377.

Shiells, C. R., R. M. Stern, and A. V. Deardorff. 1986. Estimates of substitution between imports and home goods for the United States. *Weltwirtschaftliches Archiv* 121 (No. 3).

SOFRECO (Societé Française de Réalisation, d'Etudes et de Conseil). 1988a. Etude sur la filière arachidière au Sénégal: Rapport d'avancement. Report submitted to the Republic of Senegal, Ministry of Rural Development. Paris.

_____. 1988b. Etude sur la filière arachidière au Sénégal: Rapport de deuxième phase. Report submitted to the Republic of Senegal, Ministry of Rural Development. Paris.

Stryker, D. 1990. *Trade, exchange rate, and agricultural pricing policies in Ghana.* World Bank Comparative Studies. Washington, D.C.: World Bank.

Svedberg, P. 1991. The export performance of Sub-Saharan Africa. *Economic Development and Cultural Change* 39 (April): 549-566.

UNECA/FAO (United Nations Economic Commission for Africa/Food and Agriculture Organization of the United Nations). 1985. Report on groundnut production, marketing, processing and trade in AGC member countries. Report submitted to the African Groundnut Council by UNECA/FAO, Addis Ababa.

Varian, H. 1984. *Microeconomic analysis*, 2nd ed. New York: W. W. Norton.

Watt, B. K., and A. L. Merrill. 1963. Composition of foods: Raw, processed, prepared. In *Agriculture Handbook*. Washington, D.C.: U.S. Department of Agriculture.

World Bank. 1987. *World development report 1987.* Washington, D.C.

_____. 1988. *Price prospects for major primary commodities.* Vol. 2, *Food products and fertilizers, and agricultural raw materials.* 814/88. Washington, D.C.: World Bank.

_____. 1990. *World tables, 1989-90.* Baltimore, Md. U.S.A.: Johns Hopkins University Press.

Ousmane Badiane is a research fellow at the International Food Policy Research Institute. Sambouh Kinteh is director of the Economic and Commercial Department of the African Groundnut Council, Lagos, Nigeria.